Technology and the New Generation of Active Citizens:

Emerging Research and Opportunities

Paolo Beneventi
Independent Researcher, Italy

A volume in the Advances in Public Policy and Administration (APPA) Book Series

IGI Global
DISSEMINATOR OF KNOWLEDGE

Published in the United States of America by
IGI Global
Information Science Reference (an imprint of IGI Global)
701 E. Chocolate Avenue
Hershey PA, USA 17033
Tel: 717-533-8845
Fax: 717-533-8661
E-mail: cust@igi-global.com
Web site: http://www.igi-global.com

Library of Congress Cataloging-in-Publication Data

Names: Beneventi, Paolo, 1953- author.
Title: Technology and the new generation of active citizens : emerging research and opportunities / by Paolo Beneventi.
Description: Hershey, PA : Information Science Reference, [2018]
Identifiers: LCCN 2017022527| ISBN 9781522537700 (hardcover) | ISBN 9781522537717 (ebook)
Subjects: LCSH: Political participation--Technological innovations. | Internet and activism.
Classification: LCC JF799.5 .B46 2017 | DDC 323/.042--dc23 LC record available at https://lccn.loc.gov/2017022527

This book is published in the IGI Global book series Advances in Public Policy and Administration (APPA) (ISSN: 2475-6644; eISSN: 2475-6652)

British Cataloguing in Publication Data
A Cataloguing in Publication record for this book is available from the British Library.

Advances in Public Policy and Administration (APPA) Book Series

ISSN:2475-6644
EISSN:2475-6652

Editor-in-Chief: G. David Garson, North Carolina State University, USA

MISSION

Proper management of the public sphere is necessary in order to maintain order in modern society. Research developments in the field of public policy and administration can assist in uncovering the latest tools, practices, and methodologies for governing societies around the world.

The **Advances in Public Policy and Administration (APPA) Book Series** aims to publish scholarly publications focused on topics pertaining to the governance of the public domain. APPA's focus on timely topics relating to government, public funding, politics, public safety, policy, and law enforcement is particularly relevant to academicians, government officials, and upper-level students seeking the most up-to-date research in their field.

COVERAGE

- Government
- Law enforcement
- Political Economy
- Politics
- Public Administration
- Public Funding
- Public Policy
- Resource Allocation
- Urban Planning

IGI Global is currently accepting manuscripts for publication within this series. To submit a proposal for a volume in this series, please contact our Acquisition Editors at Acquisitions@igi-global.com or visit: http://www.igi-global.com/publish/.

The Advances in Public Policy and Administration (APPA) Book Series (ISSN 2475-6644) is published by IGI Global, 701 E. Chocolate Avenue, Hershey, PA 17033-1240, USA, www.igi-global.com. This series is composed of titles available for purchase individually; each title is edited to be contextually exclusive from any other title within the series. For pricing and ordering information please visit http://www.igi-global.com/book-series/advances-public-policy-administration/97862. Postmaster: Send all address changes to above address.

Titles in this Series

For a list of additional titles in this series, please visit:
https://www.igi-global.com/book-series/advances-public-policy-administration/97862

Establishing Food Security and Alternatives to International Trade in Emerging Economies
Vasily Erokhin (Harbin Engineering University, China)
Business Science Reference • ©2018 • 406pp • H/C (ISBN: 9781522527336) • US $195.00

Smart Technologies for Emergency Response and DisasterManagement
Zhi Liu (Waseda University, Japan) and Kaoru Ota (Muroran Institute of Technology, Japan)
Information Science Reference • ©2018 • 312pp • H/C (ISBN: 9781522525752) • US $225.00

Driving Efficiency in Local Government Using a Collaborative Enterprise Architecture...
Amit Tiwary (Principal Enterprise Architect, Australia)
Information Science Reference • ©2017 • 150pp • H/C (ISBN: 9781522524076) • US $120.00

Impacts of Faith-Based Decision Making on the Individual-Level Legislative Process...
Karla Drenner (Kaplan University, USA)
Information Science Reference • ©2017 • 186pp • H/C (ISBN: 9781522523888) • US $120.00

Handbook of Research on Emerging Business Models and Managerial Strategies...
Lindy Lou West (Griffith University, Australia) and Andrew Worthington (Griffith University, Australia)
Business Science Reference • ©2017 • 509pp • H/C (ISBN: 9781522525370) • US $245.00

Ideological Messaging and the Role of Political Literature
Önder Çakırtaş (Bingol University, Turkey)
Information Science Reference • ©2017 • 317pp • H/C (ISBN: 9781522523918) • US $175.00

Therapeutic Jurisprudence and Overcoming Violence Against Women
Debarati Halder (Centre for Cyber Victim Counselling (CCVC), India & Unitedworld School of Law, India) and K. Jaishankar (Raksha Shakti University, India)
Information Science Reference • ©2017 • 344pp • H/C (ISBN: 9781522524724) • US $230.00

For an enitre list of titles in this series, please visit:
https://www.igi-global.com/book-series/advances-public-policy-administration/97862

IGI Global
DISSEMINATOR OF KNOWLEDGE

701 East Chocolate Avenue, Hershey, PA 17033, USA
Tel: 717-533-8845 x100 • Fax: 717-533-8661
E-Mail: cust@igi-global.com • www.igi-global.com

Table of Contents

Preface vi

Chapter 1
At the Dawn of Digital Revolution: Problems and Hopes of a Globalized World 1

Chapter 2
When Technology Becomes Popular: A Multimedia, Shared Production for the Information Age 23

Chapter 3
Let the Children Play! Born Into Technology, Beyond the Transmitted Knowledge 45

Chapter 4
Technology Is Not a One-Way Road: At a Crossroads – A Real Revolution or Only Market? 73

Chapter 5
Consumption and Production: Culture and Food Between Global and Local Management 101

Chapter 6
Aware Consumers Are Active Citizens Too: Participation as a Possible Strong Individual and Collective Solution 123

Related Readings 151

Index 171

Preface

Information society is where we live. There, humans learn the world from words and from images, flood of words and flood of images, more and more, flowing globally, across the groups and the nations.

Technological devices are everywhere and, though they make our lives simpler and easier from many points of view, people have generally the idea that we cannot control them and the only way is to adapt to the change. Besides, with the largest plenty of means of communication in the history, it often seems very difficult to communicate among individuals, groups, communities, countries. And technology, globally, sometimes in our perception more than in the reality of things, does not protect us from war, hunger, social conflicts, terrorism and fear.

After having been working for many decades in education and communication using every kind of technology, sometimes upstream compared to market trends, but always basing on lived experience rather than on abstract statements, in this book I try to tell that maybe other ways are possible and that we can choose, in the end, if to live in the global world as active citizens, taking part in what happens and sharing one another joys and responsibilities, or as passive consumers of goods and services, lonely and helpless in front of a world we can occasionally enjoy but never control.

The look is not only at present years or weeks, as often it seems to be when some opinion leaders and disseminators explain us "where the world is going", but it takes into account what has happened during last half a century, that it is to say more according with human times, those in which ideas, manners and behaviors of the people take shape and develop in communities, following or opposing the political, social, cultural, technological changes.

Topics and examples are picked up around, without any claim of systematic, but drawing – that is the intention – an overall framework in which the pieces of the puzzle go possible to get their place, having human experience always as the main reference.

My target audience is above all the world of education, as what we lack more is a basic literacy, in the end. School – though it is not exactly the same in different countries and situations - is going through difficult times, nowadays, in an attempt to disentangle among the educational curricula of tradition, distinguished mainly between sciences and humanities, and the pressures of the market of technology, urging the need of acquiring skills often quite ephemeral and sometimes suggesting charming definitions, as strong for the imagination as insubstantial in the reality, such as the idea of the "digital natives", which "captured the imaginations of teachers, parents, journalists, commentators and academics" (Bennett & Corrin, 2018). That is to say that we have not understood yet – as society as a whole - if we have to set up a renewed kind of education for the new generations of the digital era, or if babies today are born and grow already different, new and "automatically" able to face the future, with all another mind than their fathers, only by playing with digital devices from the youngest age. So, the teachers, do they have to teach the "digital", or to learn from the "natives"? It is not a small matter, but in recent years, at least in my country, many have played a lot and too much on this ambiguity.

And as education is not a separate planet, such clichés and gossip about "digital", conquering wide sectors of the public opinion, are having important effects on school programs, law projects, investments and expenses, on how in educational institutions the resources are oriented when the matter is to choose, for example, if spending their little money more for the renewal of the technological equipment or for the reception of immigrant children, or the assistance to kids with disabilities.

A "secondary" target is the public opinion more in general, average people moving often random between innovation and participation, in societies where social relations are more and more difficult and where because of the individual and *random* use of technology, often we do not seize some great opportunities it gives to improve the lives of all, maybe together and in cooperation, More ability to manage information and communication is what we all need, to be able to select in the huge amount of everyday news, to go and search for the useful information from the memories we are collecting in even larger amounts and we almost never go to recall, entrusting the largest part of interactions, personal but also social and political, to quick and superficial short messages and *tweets*.

To know a little the story of latest decades, and to think of it, it may be useful also to understand that anyone of us can be not only a consumer, but also an active producer of information (in the society of information), with probably more personal and collective satisfaction.

As a third target, I am addressing also to producers of hardware and software who, despite of the recurrent claimed need, for the "future", to diversify production to meet the growing custom and niche demands, have actually done this only for a few well defined professional and industrial applications, whilst for the mass of the public the proposals are, as a final result, more and more standardized, equal for all, often at the edge of fundamentalism.

This is also for lack of a true feedback from consumers, but it is the usual snake biting its tail. If, for example, during fifty years products are given for making home cinema and video, from Super 8 cartridges to analog and digital camcorders with cassettes and cards and smart phones, from home 8 mm film projectors to You Tube, and the average customers go on shooting and watching their home productions basically in the same way of fifty years before, without any idea of how a cine or video communication is to be made, it is quite a desperate job. Some features and improvements of devices match a very little advanced part of the users, but the big market has to appeal to the mass of the illiterates, who are attracted by objects beautiful and up-to-date and systematically know and use only a very small part of their power (we can think just about mobile phones), so that research, investments, production, advertising and everything are thrown as into a bottomless pit, a black hole in which all will be lost and forgotten in a short time.

The market is saturated, competition is fierce, and the opportunities which at a certain point are periodically given especially on the software side, of possible shared modes of use and production, towards a future when something new happens than the usual scenery where seller simply sell and buyers simply buy, those opportunities are not seen and considered, and are soon forgotten, if we do not get used to a more aware way of looking at all that is moving all around. Starting with knowing the history in which we live, and reflecting sometimes on it, on how the technological evolution has changed of course some parts of our lives, but also on how, despite of the current idea that all is a non-stop changing, other significant parts of our lives have changed very little, or not at all.

Not to talk about all another subject, we can see, for example, that until not many years ago, means made an impassable difference between professional producers of images and average people. Film for cinema and photographs was expensive, TV had to be broadcasted, so that, if an important event occurred, professionals went there and made a big amount of pictures and then they selected some of those pictures and gave to the public. What citizens had to do was dealing with that "digested" information and watch, listen, eventually try to understand. Sometimes citizens complained the impossibility of making themselves their own images and information.

In that context, the masters of the means held the world in their hands, ruling economy and politics, worldwide.

Today, we all make information and, if an important public event occurs, the usual crowd of cameramen taking their huge amount of pictures for a professional use is surrounded by another crowd of common people taking their own huge amount of pictures which, together with the private images taken in every other possible occasion, feed the flow of millions and billions of photographs and videos running across the social networks and filling the *cloud*, as every smartphone in our pocket can shoot with good or even professional quality.

In this context, the masters of the means still hold the world in their hands, ruling economy and politics, worldwide.

The great majority of people taking videos with a smartphone, hold it vertically. Even they take vertical panoramas, and that seems to be against any logic, having at disposal powerful wide angle lenses capable of capturing broad views of the world, but so it is.

In the mind of the *Homo sapiens* subspecies *consumator* – this is an idea of mine, but there must be a reason! - it seems to be strong the need of deeply splitting the "private" videos which we create by ourselves for our handy personal Narcissus mirror (Recalcati, 2014) and for sharing with friends on social networks from the act of watching television, possible on huge 55" or more 16:9 wide screens, and probably tomorrow on even wider ones, as screens with 21:9 (7:3) proportions begin to be on the market.

If brought into the wide screens which are common today – as it often happens when for example TV News broadcast video reports taken in real time by nonprofessionals who were on the site of the stories when professional cameramen were not – those vertical shootings fill only a narrow central part of the image, leaving all the rest black (or artificially filled, in TV studios, with color and light effects like a hazy mirror). Usually, these shootings are also very trembling and continuously moving from the left to the right and vice versa, as it is very hard to see all the action, whatever it is, in so tight a framing. Why, when it would have been enough to turn the phone to avoid all these problems?

It is as if shooting vertically and so choosing the way of the incompatibility of the format, we wanted to keep ourselves far from the responsibility of actually making real videos, that would mean to take an active part in the main flow of communication of the information society. Maybe, this is a world that we greedily consume, but where often we are not at ease and we live more and more among fears and complaints. So, if the major companies, the professionals make and sell their horizontal audiovisual products for entertainment and communication, we, apart from spending a significant part of our lives just watching their productions, have nothing to do with them!

It is maybe an interesting topic for psychologists and psychoanalysts. But apparently, in a society where everything is said to be by now always connected, the majority of people, beyond the increasingly detailed suggestions of the market (there's an app for everything now!) seem to be losing the capacity just of connecting the "different" areas of their own experience, even when those areas would in fact be very close. So, TV and our videos made with a phone are usually as separated planets, and the market encourages this sort of schizophrenia, making it normal what a few years ago would have seemed nonsensical. That's to say: the devices mainly used today for taking home videos, the smartphones, have no more possibilities of connection with a television set (unless one is a technician pretty good).

One of the aims of this book is just to point up the need, actually and not only in words, of setting up connections, and thinking of their importance and quality, in the minds of humans and in social life, as a vital condition for a world inhabited by active citizens more than by passive consumers, by human beings able and possibly pleased to meet other human beings. Connections between the technology of today and that of yesterday, as parts of the same story, minding not only at the offers from the shelves of real and virtual stores just now; connections between the sensorial world and the electronic one, possibly without all those false oppositions such as e-books against paper books, computers against body expression, outdoor games against video games, and so on; connections across the traditions and the cultures of the different peoples of the world, in times when from any personal computer, or tablet, or mobile phone we have instant access for free to automatic translations into and from every language, maps very detailed of every country, town and village of the planet, networks of webcams all across the earth, pages and groups and messages allowing all of us to hold dialogues in real time with the inhabitants of the world.

To be "connected with quality" can be something very different from the global supermarket often shown in TV phone companies' commercials. And it seems to be also rather urgent, if then in fact the same people who claim the need of the broadband all across the planet and the globalization of free economy, sometimes support the construction of walls among nations. That's to say that our hyper technological civilization does not prevent us from having just the same fears and solutions of the emperors of ancient China and Rome, one or two thousand years ago.

We can suppose that if we really get used to set up connections, if we learn to enjoy for real the inclusive side of technology, which allows all of us to talk, to meet, to know, to keep informed and to express ourselves, to produce directly a significant part of the communication, probably then we become able to face better also dangers and even enemies as they are in the reality rather than as we like to draw them in ideological visions coming straight from the ancient past.

And we can maybe also find solutions for production and work, or for the health of the planet, really adequate to times. Active citizens, connected and communicating, can begin to move together immeasurable streams of thought, in a global brainstorming where ideas, thanks to the possibilities also of organization given by the technology, much more easily could be transformed into real facts. As this is no more necessarily the old industrial society, where the old small groups of managers and politicians are the only who make decisions, with the addiction of some new problems as the competition in factories between humans and robots. Neither it is its "socialist" mirror, where the alternative to capitalism was indicated in the direction of the working class, but where just the same way of production should have been applied to a planned economy without market.

This is, or better this could be, just now, technically speaking, a society where ideas, problems and solutions can be in great part shared among all the citizens, completely changing what the world has been during thousands of years.

We have already the hardware and the software allowing us to begin to act in this way, throughout the world. We do not have yet women and men ready for it.

ORGANIZATION OF THE BOOK

Chapter 1, "At the Dawn of Digital Revolution: Problems and Hopes of a Globalized World", is a look towards the Sixties and the Seventies of last Century, where cultural and technical foretastes of the present time can be found: contradictory aspects of the welfare; young culture as a radical break with the past; possible solutions from alternative practices in movements whose protagonists are children and women and from the teachings of the great educators, which may be applied not only for school, but for the whole of society.

Chapter 2, "When Technology Becomes Popular: A Multimedia, Shared Production for the Information age", is about the spread of technology, from the second half of the Seventies, when video and computers become for all. An overview on needs, possibilities, market, ideology, with humans beginning to "think digital" and great industrial companies having sometimes to follow the suggestions of a public made not only of consumers.

In Chapter 3, "Let the Children Play: Born Into Technology, Beyond the Transmitted Knowledge", I make observations and give examples from selected experiences on how especially kids, meeting with some aspects of the recent technological evolution, can show roads and solutions good for the life of people and the active citizenship and not only for the market. It is not an easy, straightforward and already drawn way, as much depends on our ability to understand what moves under the surface and on our attitude.

Chapter 4, "Technology Is Not a One-Way Road", points out that we are as at a crossroads between "revolution" and market. The plans of big companies are not always farsighted, future is difficult to be imagined, old habits influence much our approach to the new and "emerging" trends are sometimes much related to the past, when the focus is on quantity more than quality. Innovation probably is not to be sought in what everyone does, but is scattered in many trickles, at which we should look at, more than only to mass behaviors.

Chapter 5, "Consumption and Production: Culture and Food Between Global and Local Management", is about the present situation, with hundred million people on social networks, all producers often unaware of information. Every generation – yes indeed, more the very young than the very old, but we could have funny surprises - takes part in this movement throughout the planet, though the upside-down world of technology puts often people uneasy. Maybe a chance is to restart from the humans, facing first - very significant topics as well as affecting all of us directly and in a similar way - the approach to culture and food: a big challenge.

In Chapter 6, "Aware Consumers Are Active Citizens Too: Participation as a Possible Strong Individual and Collective Solution", there is a description, by means of some examples and suggestions, of the (possible) active citizenship in the information society. A new power is in the hands of women, men and even kids, who are also called to new responsibilities. Maybe a human, active, shared approach to the technology can bring new ideas also for some sectors of the economy and lead, more than many expect, to a (at least temporary) happy ending.

Someone might be surprised as I often speak here of children as examples but, beyond the commercial and ideological use of them sometimes made on media (a huge part of the market is oriented towards children!), beyond the traditional image of them still predominant in many parts of the society, many activities with kids, and especially their answers, have shown in the years, often beforehand, unexpected possible solutions to many problems of the present era.

Something as to say, without taking it too seriously, but also not too much as a joke: "Unless you change and become like little children, you will never enter the kingdom of heaven" (Matthew 18:3).

REFERENCES

Bennett, S., & Corrin, L. (2018). From digital natives to student experiences with technology. In *Encyclopedia of Information Science and Technology* (4th ed.). Hershey, PA: IGI Global. doi:10.4018/978-1-5225-2255-3.ch219

Recalcati, M. (2014). *L'ora di Lezione: Per una erotica dell'insegnamento.* Torino: Einaudi.

Chapter 1
At the Dawn of Digital Revolution:
Problems and Hopes of a Globalized World

ABSTRACT

Our societies are technology addicted, in their deep functioning and in the daily habits of average people, who greedily consume apps and devices. A market oriented ideology is though aging things very quickly and making the memory of the past fade, even the recent one, so that many live as suspended on a nonstop present, projected towards a fanciful future, losing roots and perspective. The short period consumption landscape is also a hurdle to face as aware and active citizens the economical, social and political points of crisis, locally and globally, causing often social disruption, fear and insecurity. Going back to the years when the story began, watching diversities and analogies from the present, it can be useful to understand better our time, to deal with technology with more awareness and maybe to find out new or "old" solutions to some problems and live a better life.

DOI: 10.4018/978-1-5225-3770-0.ch001

INTRODUCTION

Pictures From the Sixties and the Seventies

Human societies are based on their history.

They are not the "trends of the market" nor the advertising campaigns, which can explain how or where the world is going to. And it is impossible for us to understand the present time, if we count out our past, from which this present comes.

Facing the technological era in which we live, in many respects, the roots we should look at are from the Sixties and the Seventies, when a strong economical and social growth alternated with inflation and unemployment crisis and many things begun, which have not been accomplished yet (Fazi, 2017).

After the World War II, consumer society has spread in North America, Western Europe, Japan, in a planet which is going to be ruled with computers and fears atomic bombs. I am presenting a quick "gallery of pictures", maybe useful to give a general idea of facts whose deep meaning is not to be explained, but takes shape in the eyes and mind of the viewer. Then I put together some elements of those pictures on a sort of time line, trying to point out connections and suggesting insights, some more "historical" and some probably enlightening nowadays too.

The subject is not simple and a lot has been written about, movies have been made, studies are still in progress.

We probably should pay attention more than we usually do to the similarities between a time that might seem far away and the present, to the fact for instance that the music of those years is still listened and played today, as well as many ideas and preconceptions about digital means and technology, if we look just below the surface, remain from those times doggedly still. Also many social problems were a clear anticipation of the problems we have to face in the present, and maybe also some possible "old" solutions could be considered, not always thinking that the world is really running at the same pace at which they tell us it goes the market of technological devices.

Picture One: The Young Music

We still listen to it and many play it, the young of all generations, as that music has not yet finished to surprise, to excite, to express the sense of the present time. Other music, of course, has been produced later and has interpreted more in details the special feeling of some moments, some social groups, but hardly we can say that it has "overcome" what rock and pops bands invented and played in the few years between the Sixties and the Seventies.

Beat music of the early Sixties is very easy, naïf and – we have to say it – not exactly always well played by very young and often poor musicians. But, though the quality is generally quite inferior compared to the best rock'n roll bands of the Fifties, it is the mark of a new freedom of expression, an actual break from the past. The public opinion discover the "young", and if mini-skirts and long haired boys are clichés which cannot be applied to all a generation, what moves around that kind of music is not only a generic rebellion to rules, nor a mere wind of nonconformity of the teen-agers, but something that affects the whole of a deeply transforming society. That music is the audible voice, the spearhead of the change.

There are also though some "never-beat" groups, such as The Rolling Stones and The Who, and in the late Sixties the quality of young music improves incredibly. Only a few names, I quote by heart: The Doors, The Cream, The Pink Floyd, Procol Harum, Santana, Traffic, Jefferson Airplane, Deep Purple, Led Zeppelin. Jimi Hendrix re-invents the electric guitar, and the same Beatles, in the second part of their as short as extremely rich career, deeply evolve.

In the early Seventies the level of young music grows remarkably high, in some case very sophisticated and complex: Jethro Tull, Genesis, Yes, Emerson Lake & Palmer, King Crimson, The Queen. It is the perfect sound track for the social mass movements which spread worldwide, mixing culture and instincts, peace and violence, hope and despair, bringing massive new behaviors and values in new generations of men and women. Still often very young, but now also very good groups and musicians appear also out of the dominant context of the Anglo-Saxon countries. If Italian beat music – an example from my country - was considerably inferior to the English and the American ones, now groups as Premiata Forneria Marconi, Banco del Mutuo Soccorso, Area can stand proudly on the international scene.

Music is a true, very important expression of those times, of the hopes and the conflicts, the ideas of changing the world in which a great part of young people is engaged. And probably as it is a deeper, artistic expression of the human soul, it happens that, while many of the ideologies of the Seventies are probably over forever, the music of those years is still here, alive, listened and loved today by generations of grand-fathers and grand-sons. And a thing like this does not happen by chance!

Picture Two: The Student Revolts

The student occupation of Sproul Hall on December 2, 1964 (...) was in response to the proposed expulsion of Jack Weinberg and other members of the Free Speech Movement and other student political groups for their involvement in the recent protests on the Berkeley campus, as well as for the Universities refusal to drop charges against student political group leaders. The initial plan was to occupy the hall for a single night; however, the protesters were prepared for a two to three-day demonstration and siege of Sproul Hall. Ultimately, 1500 students occupied Sproul Hall before being removed by police. In all, 773 student activists were arrested for their involvement in this event. (1960s Berkeley protests, 2017)

In many historical periods students have been engaged in political movements, but nothing probably ever happened like in the Sixties of last century, with so large a participation, worldwide, and so great an impact on the civil society. From California, within few years, the movement spread in many capitalistic countries, above all in Germany, Italy, England, France, and also in socialist countries, as Czechoslovakia, Poland and Yugoslavia. The most significant year is 1968, when very dramatic events occur, as the Tlatelolco massacre of students in Mexico City, with at least 27, officially, but maybe, according to other sources, up to 300 killed (Doyle, dateless), or the end of the so called "Prague Spring" (not only a student movement, of course), repressed by the intervention of tanks from the Soviet Union (with sporadic acts of violence and several suicides). But the "symbol" of all that season is the French May, when to the demonstrations of students massive general strikes of the workers follow, as well as the occupation of universities and factories.

It was a year of seismic social and political change across the globe. From the burgeoning anti-Vietnam war and civil rights movements in the United States, protests and revolutions in Europe and the first comprehensive coverage of war and resultant famine in Africa. The world would never be the same again. (The Guardian on line, 17 Jan 2008)

Picture Three: From Technology, a New Possible Interactivity

I like to quote Roberto Maragliano when he writes of the different kinds of knowledge induced by the different means through which knowledge itself develops in human minds. The book oriented culture follows "the paradigm of the text … knowledge as a physical object"; TV culture "the paradigm of flow… knowledge as immersive environment" and what we do with digital devices follows "the paradigm of interactivity" (Maragliano, 2007).

I like also to say that interactivity can be a quite poor approach if not completed with the other great possibility given by the digital, that is *sharing*, with an eventual change of roles between producers and consumers, but that is in the end one of the main subjects of all this book, so here I only consider the three ways of approach to a given culture meanwhile, not yet transforming: text, flow and interactivity. To point out that the time of interactivity begins in the Seventies of last century, with the invention of home video recording and personal computers.

More details about the technical and commercial aspects of the entry of technology in the daily life of all will be better developed in next chapter. Here it is useful to remember that the same generation just coming out from the '68 begins to play video games in bars and to take moving pictures using the first video tape recorders. May we say that this can mean somehow, in the right hands and with the right intentions, to give power to the imagination? (Von Dirke, 2016)

Picture Four: The Women's Liberation Movement

In latest times, another reason of worry in the world - as well as economical crisis, unemployment, wars, terrorism, uncontrollable flows of migration, environmental disasters - is a general increase of violence against women.

According to the Online Encyclopedia of Mass Violence:

[Femicide] is proposed as an alternative to the gender-neutral term of 'homicide'. As such, it seeks to highlight the killing of women for being women. … . Femicide is distinct from general homicide because the instances occur in domestic settings as a result of intimate partner or family violence. … Violence against women constitutes the most common crime in the world with the highest levels of impunity for perpetrators. (Femicide, 2017)

I know that this is not probably the most optimistic way of introducing the women's liberation movement which, in the Seventies, in fact began to change a significant piece of the world. But this is another great social issue strongly emerged in those years, whose failure to complete evidently we pay dearly today. Or better, we should always remember that every social achievement or civil coexistence rule can never be taken for ever, that also supreme values as democracy and peace cannot be a heritage of the past to be enjoyed, but must be built day by day, by active citizens who engage in person to save them.

Women movement of the ’70 is the second great wave of *feminism*, after that of the beginning of 1900 for the right of women’s suffrage, and stands as a strong critics to the traditional roles in family and society as well as to other alternative movements such as those of students and workers, which claim to revolutionize the world without solving within them the contradiction between man and woman.

“The personal is political!” is a typical rallying slogan (Hanisch, 1970 – 2006), affecting also deeply the student and other left movements in which women are involved. In Italy, where in few years a deeply catholic society evolves from a late introduction of divorce to an advanced law on free (also of charges) right of abortion, other slogans are “I am mine!”, “The uterus is mine and I manage it!!”, “Neither whores nor madonnas, only women! !” (in Italian, of course with the appropriate rhyme!) and women seem to be the protagonist of the reinvention of the political (Bracke, 2014).

In those years, not without problems and contradictions, women were actually conquering a new freedom day by day, above all affirming the right to a sexuality not linked to predetermined roles. Some behaviors which before were rare and rebellious become massive, and the whole of the society becomes to get used with them. This also applies to a gradual acceptance, as not an abnormal human condition, of homosexuality.

So, I totally disagree with those people who, commenting violence on women from present daily chronicles, say that it is because men hardly accept a society that is quickly changing. Probably the reason is just the opposite: men are more violent just because the society has stopped changing.

Beyond the Welfare: The One-Dimensional Man Fears the Future

Some facts can be taken, to represent a rich and contradictory period, facts not necessarily real, as from cinema, TV, music, literature and papers ideas capture the imagination of people, sometimes stronger than real experience.

The TV series *The Jetsons* are cartoons for kids by Hanna-Barbera, airing first between 1962 and 1963. They are based on a utopian vision of the future, when it will be enough to press a button to make things happen. A very strong idea, actually much rooted in the mind of average people, the same winning idea which, after the weird time of personal computers, so little automatic and so asking active (and from a certain point of view annoying) human actions and instructions (not for all indeed, not the ideal for best selling on the market), will be restored in the reality of the new millennium, when "technology" easy and cheap, ready for use, will be really given to everybody at last, in the form of mobile phones and social networks.

In January 1963 the Beatles sing *Please Please Me*. A juvenile wave is mounting throughout the world, not only music, but habits, morals, which in some cases are strongly opposing the dominant values. Recreational drugs as cannabis and LSD are normally used in most of the young movements, in the so called *underground culture* and among the *hippies*, which in the second Sixties also refuse welfare and consumerism:

Hippies felt alienated from middle-class society, which they saw as dominated by materialism and repression. . . . Hippies advocated nonviolence and love, a popular phrase being "Make love, not war," for which they were sometimes called "flower children." They promoted openness and tolerance as alternatives to the restrictions and regimentation they saw in middle-class society. Hippies often practiced open sexual relationships and lived in various types of family groups. They commonly sought spiritual guidance from sources outside the Judeo-Christian tradition, particularly Buddhism and other Eastern religions, and sometimes in various combinations. Astrology was popular, and the period was often referred to as the Age of Aquarius. (Hippie, 2017)

Music was an integral part of all the alternative movements, as well as a strong element of contradiction, as it was at the same time, on the other side, a multi-millionaire business.

Herbert Marcuse's book *One-dimensional Man* comes out in 1964. The author "offers a wide-ranging critique of both contemporary capitalism and the communist society of the Soviet Union, documenting the parallel rise of new forms of social repression in both these societies, as well as the decline of revolutionary potential in the West" (*One-Dimensional Man, 2017)*.

The book opens with these words:

A comfortable, smooth, reasonable, democratic unfreedom prevails in advanced industrial civilization, a token of technical progress. Indeed, what could be more rational than the suppression of individuality in the mechanization of socially necessary but painful performances; the concentration of individual enterprises in more effective, more productive corporations; the regulation of free competition among unequally equipped economic subjects; the curtailment of prerogatives and national sovereignties which impede the international organization of resources. . . . The rights and liberties . . . freedom of thought, speech, and conscience were - just as free enterprise, which they served to promote and protect - essentially critical ideas. . . . Once institutionalized, these rights and liberties shared the fate of the society of which they had become an integral part. The achievement cancels the premises. (Marcuse, 1964)

Marcuse's ideas are going to influence strongly the student revolts of the second half of the Sixties, together with the ideas of many others, philosophers who criticize the consumerist society as well as theorists of the revolution, traditional communists and anarchists in all their versions (in Europe multiply the Maoist movements, often one against the other) as well as members of anti-imperialist and civil rights liberation movements.

In October 1967, while trying to foment a revolution with his guerrilla force, Ernesto Che Guevara is killed by the Bolivian army:

He became a myth in his own lifetime and an international martyr-figure upon his death. . . . Che Guevara's dream was an epic one - to unite Latin America and the rest of the developing world through armed revolution and to end once and for all the poverty, injustice and petty nationalisms that had bled it for centuries. (Che Guevara: A Revolutionary Life, 2017; Anderson, 1997 - 2010)

In Italy, to the '68 the '69 follows, whose protagonist are the workers of industry in the north of the country, during the so called Hot Autumn. In the USA, also embroiling peaceful movements as the aforementioned hippies or others much less peaceful as the Black Panthers, a large mass protest is growing against the Vietnam War.

Understanding Media: The Extensions of Man, had been published by Marshall McLuhan in 1964, the same year of Marcuse's book, so it begins Chapter 1:

In a culture like ours, long accustomed to splitting and dividing all things as a means of control, it is sometimes a bit of a shock to be reminded that, in operational and practical fact, the medium is the message. This is merely to say that the personal and social consequences of any medium - that is, of any extension of ourselves - result from the new scale that is introduced into our affairs by each extension of ourselves, or by any new technology. (McLuhan, 1964)

I remember when distinguished scholars quoted with skepticism the words of McLuhan, but more than fifty years after, looking at what we can see around, with the latest objects of technological desire, probably they are worth to be rethought and meditated.

The film *2001: A Space Odyssey* is out in halls in 1968. The story is very optimistic about the development of space exploration, telling about large space stations, permanent bases on the Moon, powerful space ships in mission up to Jupiter, but quite pessimistic about computers, in the "future" imagined so big and so possibly out of control and dangerous.

A year later, real men move their first steps on the surface of the real Moon, but the way of space is uncertain and too much expensive, so that space ships will continue until today and for much time ahead to fly above all in our dreams and imagination.

On the other side, only a few years later, the invention of the micro-processor is going to change completely the development of computers and digital devices, so that a fiction machine like HAL in Kubrick's movie, already in the middle of the Eighties will appear irremediably outdated.

But ideas, as we have already remarked, are not always based on the actual reality, and what was represented in the rebel super computer of Kubrick's space vessel is maybe a quite recent, but deep fear of our times.

Many science fiction writers, since the Fifties, have imagined a future of humans who enslave other humans "automatically" using machines, or even of intelligent machines which overcome totally the humans and enslave all of them according to an internal logic of their, utterly incomprehensible and inevitable, no doubt ruthless. And computers and robots, as any entity we do not know directly and so we do not understand, have been a cause of fear well before average people could face at last the real ones, when computers became personal. Common people - since the first computer on the market, the Univac, was accepted by the United States Census Bureau on March 31, 1951, for many years had only listened of computers, as electronic brains, hide in the rooms of the army, the universities, the data and research centers, public administration, and their idea of them, a strong idea, had been taking shape from vague news from newspapers, magazines, TV, and above all from science fiction books and films.

Computers, according to the current imagination, should have been something like in *The Jetsons* (the positive, frivolous side) or in *A Space Odyssey*, (the worrying one), but also as in the *Star Trek* series: pushing buttons, speaking to intelligent machines, fearing their possible rebellion, as they appeared destined to become much cleverer than humans, able to know everything, not conditioned by emotion, hate or love.

This is probably to be remembered also in the time when computers and electronic devices have become products for a mass, huge, global consumption, as maybe human minds are not so fast as the market in "updating".

Differences and Similarities With the Present Time

If we look at the video reportages of pop music concerts, we often see crowds of young girls into a frenzy storming around the rock stars, in a very similar way in the Sixties as today. Is it a matter of female nature or that comes from something related to the society? Does it mean that our times and the Sixties, for some aspects, are closer than we usually believe? And why, in my memory of the many concerts I attended to during the Seventies, I do not remember anything similar? Were they a different kind of concerts, did we attend to them with another attitude? Can we put together and compare, for example, the Beatles at Washington Coliseum and Woodstock?

I am not a sociologist, but sociology is not an exact science and some ideas can be put on the table, if they suggest us some way of understanding better the complexity of human behaviors.

First: the "difference between generations". It is almost a natural fact in the mind of many, when we often read or watch on mass media, or when we chat in social networks about that "problem". Young are far anyhow from the adults, unknowable, unattainable, perhaps mysterious, especially when for them strong definitions are used, such as "beat generation", "punks", "millennials", "digital natives", which may not mean anything, but divide, turn away, make significant and important what in the reality can be only superficial attributes, as leather jackets, Mohawk colored hairstyle, exceeding familiarity with mobile devices. Usually, only to speak with them and above all to listen to those "aliens" would be enough to break down every distance, but maybe it is too easy and for some reason we prefer to go on thinking that generation divide is an unchanging and cyclically repeating fact, across years and decades, somehow probably required by the human condition itself.

If it concerns our own sons and daughters, or our students, as we are the parents or the teachers of those young, but also if we are public administrators who provide social policies, to know if actually a break between generation is necessary and exists, if it is always the same or is changing in time and space and living conditions, it can be quite important, unless we do not want to base our actions only on stereotypes, or on the imagine which of real human beings are given by media. And for example just the very recent years, at the beginning of the millennium, are described by several authors as a period when the generation gap is, in the reality, at its minimum. (Giancola, 2006)

Sure, also in the Tens of 2000 teen agers in their groups behave in another way from their parents (the way they dress, play, live and love) and probably the mind of mature and young people work somehow differently when they both spend much of their time on *Facebook*. In my opinion, all that is not enough to confirm the existence of a real generation gap, as there are also many points of contact, much more than in other times, as the already quoted music, the common addiction to digital devices, the way as many adults dress, disguised as young, or as many young face relationship problems, sometimes with a mentality that to their parents, when they were young in their turn, would have seemed “old”.

Though never we own the truth, dealing with social processes, and always it is present the risk of taking blunders, so that we have to think well of what we see at the surface of things, not to generalize too much, and ask the right questions rather than immediately claim to give the right answers, I think we can say for sure - after all that we have written before - that during the Sixties and Seventies of the last century the generation gap was much stronger than today, and the young people of that period were very near to express what can be called a generational culture, not only as a reflection of market trends or questionable points of view.

In fact not all the young, not every singular young, but really many young, maybe the majority, were going together much beyond the groups of friends, the fans of musicians, the supporters of sport squads, taking parts in true mass movements, with the aim and the hope of changing the world. Movements were against the “dark sides” of a still growing global consumer society, which at those times were possible more visible, as often it happens at the beginning of something. Whilst today, maybe because we are just in the middle of the storm, despite of the heavy and recurring crises, no credible alternatives seem to appear to the dictatorship of the market.

Those dark sides, just in that musical fields which we have seen connecting generations, were strongly symbolized at those times by the unexpected death of many rock stars, as Jimi Hendrix, Janis Joplin, Jim Morrison, Brian Jones of the Rolling Stones (but in more recent years, the "curse of 27" hit also Amy Winehouse, and other older stars died in dubious circumstances, as Michael Jackson, Prince, George Michael).

In the Seventies the world had to deal with many economical troubles, the oil crisis of 1973, and also a new mix of recession and high prices rising, for which a new term was invented: "stagflation" (stagnation + inflation).

The turning point was probably 1977.

In Italy, in the middle of a great unemployment crisis, there were the last, creative and desperate big student riots, at the beginning of the so called "Years of Lead", swinging between the pacific performances of "Metropolitan Indians"

Figure 1. Unemployment and inflation trade-off: at the end of the 70, high inflation coincides with growing unemployment
Source: Pettinger, 2017

and the sympathy of some little but significant sectors of the movement for the armed struggle of Red Brigades.

In the rest of the western world, the "punk" generation was a significant minority of young people who did not expect anything more from the future and dug for themselves a niche of surviving, while a general return occurred to private and individual solutions, trivial or dramatic, as traditional family and rat race, or waiver and drug addiction.

It seemed to be the end of a dream, the global failure of social and utopian movements, and the return to the rules (or no rules: the magic word is *deregulation*) of a capitalism now without opposition, which would culminate in few years in the fall of the Soviet Union.

We always well have to remember, though, that never social processes are univocal and linear. In fact, in the same year 1977 another "revolution" was announced, with the first commercial success of a new sort of machines, the so called "personal computers", whose name were *Apple 2, Commodore PET, Tandy Radio Shack TRS-80…*

At School With the Body! Children Teach Solutions for the Past and the Present

To go inside the facts means not merely to describe them, but to take the responsibility of choosing in the heap those which are significant. That is not an objective point of view, of course, but it is what can help people to watch beyond the appearances, maybe to get an idea of things and move better across the reality. Different points of view sometimes are useful to approach a topic from its different sides, as the world is not one-dimensional, as probably Herbert Marcuse would say.

A strong answer to the many questions asked during the Seventies, from the point of view not only of education, may be summarized in the title of a booklet describing several experience of what in Italy we called *animazione teatrale (*Beneventi, 1994; Beneventi and Conati, 2010) That is to say free expression, creativity, non-authoritarian education, mainly using with children and giving them the languages that lead to the theatre. That title of the booklet, simple and essential, was *At School with the Body.*

Many children enter in the school and learn mechanically, without recovery of their early experiences, which are lost to communication and sometimes lost altogether. . . . They learn something and lose it right away; they have no mechanism to incorporate what is presented to them. . . . The school, with its pretensions in precise program and prearranged times, is an immediate assault on children, who are forced to defend themselves reacting in some way: accepting programs and giving up their language and their culture: trying to be outlaws and continuing to live their language and their culture in hiding. (Canevaro, 1974, loosely translated by Paolo Beneventi)

Sure, all the critique of the bourgeois society, the cultural industry, the classist aspects in school, as well as the organization of the factory, developed from the young, students, workers movements, hardly fit in a slogan of few words as "at school with the body". But the message behind was possible stronger.

I try to explain. Movements are made of women and men, who take part in them with all their histories, ideas and ideals, education, behaviors, habits, and personal contradictions. In those years, for example, some feminists said that it is difficult to make a revolution in the name of freedom when the leaders are sexually repressed, as many in '68 were told to be (Ross, 2008). But also many feminists, after their Roaring Seventies, changed their mind and life, back to much more traditional, average, even commercial values.

Something like that often happens when we humans move – let me say so - more with mind than with heart, when we think we can change the world without taking account of the problems we have to stand in our own lives, which easily affect our thinking and acting.

Without falling in a general psychological interpretation of revolutions, once in a while we rely on common sense to say that if for example the ideas and the actions of a leader are moved – as sometimes they are – also from desire for revenge and personal frustration, probably he is not a good leader and that revolution is not destined for a durable success.

The movements of the Sixties and Seventies of the past century were often very radical, but also very ideological, and *trendy*! Many people really believed in them, but many probably were only following the stream. And – another thing that often people say - it is difficult to change the institutions if the individuals do not change. But, as individual changes are not only a matter of will, here we are back to our initial statement, to those few words.

To go to school with the body is not to be intended, as one might think, as something concerning the amount of physical education in the curriculum (though physical education is important, indeed, for a correct body development and also to fix in kids the damage of a sedentary lifestyle, obesity and other problems), but it refers directly to the typical way of very young babies of growing and knowing the world, using not only their mind, but all the senses, touching and tasting things, having tactile and motor experiences that are connected directly to language learning and to the development of thought.

The functions to be established by the child fall into two groups: (1) the motor functions by which he is to secure his balance and learn to walk, and to coordinate his movements; (2) the sensory functions through which, receiving sensations from his environment, he lays the foundations of his intelligence by a continual exercise of observation, comparison and judgment. In this way he gradually comes to be acquainted with his environment and to develop his intelligence. At the same time he is learning a language, and he is faced not only with the motor difficulties of articulation, sounds and words, but also with the difficulty of gaining an intelligent understanding of names and of the syntactical composition of the language. (Montessori, 2009)

Such a way of learning works usually well until babies are free to grow up if not alone, out of a formal instruction, but at some point, when kids have entered the school, it is suddenly replaced by the transmission of knowledge from teachers, a given and coded knowledge. The focus is shifted from collecting and linking the data of one's own direct experience to memorizing notions from external sources, and the harmonious balance between body and mind often breaks. Many children hardly accept to stay sitting among desks and working in that standardized a way, and problems start.

To be at school with the body – in the way as it happens for example in many experiences and workshops of *animazione* and more in general in authentic situations of active school - means to give the possibility to the young kids to go on growing and developing their personality adding new abilities, including the logical thought of the adults and the learning of the disciplines, without necessarily giving up forever their emotional and physical approach to reality. They can continue to follow their senses, to feed their natural curiosity and amuse, testing the power of imagination and also discovering the strength of the cooperation with the others, producing culture as well as learning, without ignoring the conflicts but going beyond them.

It is not sure to change the world yet, but it shows a possible way, an alternative model good also for a part of the society, in which the personal and public and politic areas are possibly connected (not dissociated as often they are) and the human beings are allowed, maybe, to find a sort of harmony with themselves and the others, according to their nature.

The Time Has Come to Put Into Practice the Teachings of the Great Educators

Modern society breed kids of consumers, between welfare and conformism, but one-dimensional humans often feel they lack other, essential dimensions, and from some vantage points – universities, factories – at some time, as between the Sixties and the Seventies, they try to say no, to give "All Power to the Imagination!"

I was 14 years old in '68, and I remember my first "political" act. After school time, once we gathered to a site where older students began to make us to read a thin, humble white book: *Lettera a una professoressa* (A letter to a teacher), written by the students of a small "private" school, more than popular, managed by a priest, don Lorenzo Milani, a school attended by those who were in trouble with "official" schools. They expressed a simple point of view, which sounded to us as a strong indictment of the educational institutions and the usual way of acting of teachers:

During class works, you moved back and forth among the desks and saw me in trouble or wrong and said nothing. . . . You are paid to help me and instead waste time to watch me like a thief. . . . Barbiana, when I arrived, did not seem a school. There was not a teaching post nor the blackboard, nor desks, only large tables around which we made school and ate. . . . The oldest of those teachers was sixteen years old, the youngest twelve and filled me with admiration. I decided from the first day that I would have taught as well. (Scuola di Barbiana, 1967, loosely translated by Paolo Beneventi)

In the abstract of an ebook on Don Milani recently published, in English, by an Italian house (Giacomini and Costanzi, 2016), we read: "*Le coscienze oggi sono appiattite – dice il grande pedagogista – perché il modello culturale dominante opera sul tempo libero e sulla diffusione dei media, anziché sulla conoscenza e sullo studio* [Consciences today are flattened - the great pedagogue says - because the dominant cultural model operates on leisure time and on the spread of media, rather than on knowledge and study].

This he said in 1967.

Many other great educators during the past two centuries have paid attention on how actually the children grow, learn, become aware of the world. And their descriptions were often far away from the social and scholastic models.

But the main mission of the school institutions is not, in the end, to help kids to grow according to their nature, not to follow the Constitutions, when they speak of the happiness of the citizens. School has to prepare the young to the real world, the social relationships, the production and the market. And often the hardest schools are the best, not for everybody, as the ruling class is to be formed through a strong selection.

This contradiction between the right to education for all and the selective nature of school is one of the causes of the student riots of 1968. And the possibilities of a more harmonic development of the human being coming from the descriptions made by the great educators and researcher on childhood as Montessori, Dewey, Freinet, Piaget and many others, as well as the dreams of the students who claimed the power to imagination, as well as the true experiences of the children in the classes of the active schools and in the workshops of *animazione*, all this suggests that maybe an alternative to the one-dimensional society is possible.

From the book of John Dewey *Democracy and Education*, it can be interesting to quote here the whole "summary" of Chapter 1:

It is the very nature of life to strive to continue in being. Since this continuance can be secured only by constant renewals, life is a self-renewing process. What nutrition and reproduction are to physiological life, education is to social life. This education consists primarily in transmission through communication. Communication is a process of sharing experience till it becomes a common possession. It modifies the disposition of both the parties who partake in it. That the ulterior significance of every mode of human association lies in the contribution which it makes to the improvement of the quality of experience is a fact most easily recognized in dealing with the immature. That is to say, while every social arrangement is educative in effect, the educative effect first becomes an important part of the purpose of the association in connection with the association of the older with the younger. As societies become more complex in structure and resources, the need of formal or intentional teaching and learning increases. As formal teaching and training grow in extent, there is the danger of creating an undesirable split between the experience gained in more direct associations and what is acquired in school. This danger was never greater than at the present time, on account of the rapid growth in the last few centuries of knowledge and technical modes of skill. (Dewey, 1916)

More precisely:

There is the standing danger that the material of formal instruction will be merely the subject matter of the schools, isolated from the subject matter of life - experience. The permanent social interests are likely to be lost from view. Those which have not been carried over into the structure of social life, but which remain largely matters of technical information expressed in symbols, are made conspicuous in schools. Thus we reach the ordinary notion of education: the notion which ignores its social necessity and its identity with all human association that affects conscious life, and which identifies it with imparting information about remote matters and the conveying of learning through verbal signs: the acquisition of literacy. (Dewey, 1916)

Many of most advanced works and experiences in education, since the Sixties till now, are inspired by the constructivist theory, whose founder was Jean Piaget. And an eminent follower of Piaget was Seymour Papert, the inventor of the programming language LOGO.

Piaget was one of the first to research scientifically how the children *think*:

Child logic is a subject of infinite complexity, bristling with problems at every point, problems of functional and structural psychology, problems of logic and even of epistemology. It is no easy matter to hold fast to the thread of consistency throughout this labyrinth, and to achieve a systematic exclusion of all problems not connected with psychology. If we try too soon to give a deductive exposition of experimental results, there is always the risk of succumbing to preconceived ideas. . . . The logic of the text-books and the naïve realism of common sense are both in this respect fatal to any sane psychology of cognition, and the more so since in trying to avoid the one we are often thrown back upon the other. (Piaget, 1923)

During the Seventies, some authors in France begin to speak of Psychomotor education (*psychomotricité*). The words about it from Wikipedia (though the English entry is very poor and "may not meet Wikipedia's general notability guideline"), are good enough for giving a concise idea of it:

A pedagogic and therapeutic approach, the aim of which is to support and aid an individual's personal development. It is based on a holistic view of human beings that considers each individual as a unity of physical, emotional and cognitive actualities, which interact with each other and the surrounding social environment. (Psychomotor Education, 2017)

Psychomotor education comes directly from the experience. Playing with the perception of *contrasts* as basics, with the *associations of contrasts*, structures and rhythms, and finally with *shades* (Lapierre and Aucouturier 1974 – 1975) children become aware in a fine and amusing way of their own body and the world around them.

In some countries, as Italy, this kind of education was widely taken in schools in order to monitor together the physical and mental development of young kids, as well as to correct eventual problems encountered.

If it could be really possible, or desirable, starting from education, to change the traditional competitive model of society where from a selection hard and often unnatural few emerge from the mass, into another one, where people grow in harmony and put together and share skills and knowledge, not all agree. And sure, technically, the world of the Seventies was not ready for such a change yet.

But just when dreams, utopias and revolutionary ravings appear to be over, maybe forever, technological development is going to give the people, all the people, in their own hands, tools and possibilities which never mankind has known in the history, to possibly change the world.

This comes through a radical break from the past, as coinciding with the ideas of great educators, as quite far from the common sense, also that is behind revolutions, usually made by mass movements following ideal models of society and charismatic leaders and with the individual stories kept in background and in the end not relevant, before the greatness of the big transformations. That is probably the contradiction which makes revolutions often to fail, as they do not match the real life of the women and men actually involved, and real people have to conform to "ideas", that's to say obeying to what the leaders decide.

Technology is now going to give tools available for individuals and small groups, suitable in every little, limited and local situation, but at the same time open and linkable to the global world. Technology is now going to give a new power in the end of everyone of us, as well as new responsibilities.

Responsibility means the ability of a person to act consciously and take into account the consequences of his or her actions. A person acting responsibly is capable of understanding the needs of other people. In addition, the sense of responsibility gives a person a sense of empowerment, a sense of possibility to influence the state of affairs and control over certain events in the external world. (Platonova, 2011)

There can be problems about all this, as common people is not accustomed to deal with power, and general refuse responsibilities.

REFERENCES

1960s . Berkeley Protests. (2017). Retrieved September 14, 2017 from https://en.wikipedia.org/wiki/1960s_Berkeley_protests

Canevaro, A. (1974). Il Corpo come Linguaggio. In F. Alfieri & A. Canevaro (Eds.), *A Scuola con il Corpo*. Firenze: La Nuova Italia.

Che Guevara: A Revolutionary Life. (2017). Retrieved on September 14, 2017 from https://www.amazon.co.uk/Che-Guevara-Revolutionary-Jon-Anderson/dp/0553406647

Dewey, J. (1916). *Democracy and Education. ManyBooks.net.*

Femicide. (2017). In *Wikipedia.* Retrieved on September 14, 2017 from https https://en.wikipedia.org/wiki/Femicide

Giacomini, A., Costanzi, G., & Vignati, V. (2016). *Don Milani. The Teaching Innovator*. Roma: Armando.

Giancola, F. (2006). The Generation Gap: More Myth than Reality. *Human Resource Planning*, *24*, 4.

Hippie. (2017). In *Britannica.* Retrieved on September 14, 2017 from https://www.britannica.com/topic/hippie

Maragliano, R. (2007). *Immaginare l'Infanzia*. Roma: Anicia.

Marcuse, H. (1964). *One-Dimensional Man: Studies in the Ideology of Advanced Industrial Society*. Boston: Beacon Press.

McLuhan, M. (1964). *Understanding Media: The Extensions of Man*. New York: McGraw-Hill.

Montessori, M. (2009). *Dr. Montessori's Own Handbook.* New York: Frederick A. Stokes Company Publishers. Ebook.

One-Dimensional Man. (2017). In *Wikipedia.* Retrieved on September 14, 2017 from https://en.wikipedia.org/wiki/One-Dimensional_Man

Pettinger, T. (2017). *Trade off between unemployment and inflation.* Retrieved from http://www.economicshelp.org/blog/571/unemployment/trade-off-between-unemployment-and-inflation/

Piaget, J. (1959). *The Language and Thought of the Child* (3rd ed.). London: Routledge.

Platonova, A. (2011). Responsibility in Technological Civilization: In Search of the Responsible Subject. Wien: Das Institut für die Wissenschaften vom Menschen.

Psychomotor Education. (2017). Retrieved on September 14, 2017 from https://en.wikipedia.org/wiki/Psychomotor_education

Scuola di Barbiana. (1967). *Lettera a una Professoressa.* Firenze: Libreria Editrice Fiorentina.

ADDITIONAL READING

Anderson, J.L. (1997 – 2010). *Che Guevara: A Revolutionary Life.* New York: Grove Atlantic.

Beneventi, P. (1994). *Introduzione alla Storia del Teatro-ragazzi.* Firenze: Ponte alle Grazie.

Beneventi, P., & Conati, D. (2010). *Nuova Guida di Animazione Teatrale.* Casale Monferrato: Sonda.

Bracke, M. A. (2014). *Women and the Reinvention of the Political: Feminism in Italy, 1968-1983.* Abingdon, UK: Taylor & Francis Ltd.

Fazi, T. (2017). *Una crisi iniziata quarant'anni fa. "MicroMega" 4.* Almanacco di Economia.

Lapierre, A., & Aucouturier, B. (1974). *Le Contrastes, e la découverte des notions fondamentales.* Paris: Doin.

Lapierre, A., & Aucouturier, B. (1974). *Associations de Contrastes, structures et rythmes*. Paris: Doin.

Lapierre, A., & Aucouturier, B. (1975). *Le Nuances*. Paris: Doin.

Ross, K. (2008). *May '68 and Its Afterlives*. Chicago: University of Chicago Press.

Von Dirke, S. (2016). *All Power to the Imagination!* Lincoln, NE: U of Nebraska Press.

Chapter 2
When Technology Becomes Popular:
A Multimedia, Shared Production for the Information Age

ABSTRACT

The story of mass technology begins in the Seventies of last century, when computers and video devices became "personal" and began to spread in homes. It is not only a matter of industrial production and market, as many "inventions" came from the base, common people, young students, cultural communities, sometimes driving the innovation beyond and even against the plans of the big companies. To know the story is also a way to become aware that anyone of us, possibly together, can do much more than what we generally expect, and also that "technology" is not an endless list of possibilities at our disposal more and more but, if we do not take part actively with our suggestion and choices, there is the tangible risk to waste in the swirl of market a great part of it.

DOI: 10.4018/978-1-5225-3770-0.ch002

INTRODUCTION

Common People and Technological Equipment

Here, being not a technician, but a researcher on the field in education, I tell of some machines met during the years and of their possible use, from my personal point of view, that is partial, as every other point of view, but maybe helps to speak not too much in general, to link ideas and opinions to the reality more than to ideology, which is often a big, annoying problem.

I give examples of the programming languages which every PC user in the first Eighties had to deal with, that is maybe useful not equivocate every time that we use the word "digital", and also to remember how the electronic devices and all that moves around them are not, or not inevitably, consumer products among many others. Users can command them, individually, but also, sometimes, change their overall story.

A great thing today is that, writing a book like this about facts and weird objects probably unknown by the most, we do not have to fill the pages with notes, as anyone, if likes, has the possibility to go and get in few seconds, only writing some names, a lot of news and pictures on line!

So wrote Ivan Illich, in 1978:

Modernized poverty appears when the intensity of market dependence reaches a certain threshold. Subjectively, it is the experience of frustrating affluence which occurs in persons mutilated by their overwhelming reliance on the riches of industrial productivity. Simply, it deprives those affected by it of their freedom and power to act autonomously, to live creatively; it confines them to survival through being plugged into market relations. And precisely because this new impotence is so deeply experienced, it is with difficulty expressed. We are the witnesses of a barely perceptible transformation in ordinary language by which verbs that formerly designated satisfying actions are replaced by nouns that denote packages designed for passive consumption only: for example, "to learn" becomes "acquisition of credits." A profound change in individual and social self-images is here reflected. (Illich, 1978)

This is a significant part of that "dark side" of the information society we have already mentioned. But what PCs and home video cameras showed, between the Seventies and the Eighties, is that other directions are possible.

Here I am writing often using examples, trying to point out some facts which can be significant. It is not "the reality", but some aspects of the reality which, in my opinion, are worthy of a reflection, comparison, perhaps analysis, if we want not only to certify, but possibly to understand what is happening in the world, and above all to see if and how we can take part actively in it. As it may be very frustrating to believe we have no other choice but standing by the window and looking at all that blooms and conditions our lives heavily, all so bigger and so too far away from us and from any chance of direct action.

Act 1: From Television to Home Video

Senza chiedere il Permesso, come rivoluzionare l'informazione (Without asking permission, how to revolutionize the information) by the film director Roberto Faenza, was a booklet out in Italy in 1973. The picture on the cover showed long haired young people with a camera in the hands, a video tape recorder hanging from a shoulder and a submachine gun from the other.

The point of view was quite peripheral, as in Italy in those times –only one image to describe the situation - there were not private and commercial media yet (they began to spread a few years later), but only a state radio and television. The idea though was very strong: having at disposal portable and cheap enough devices, now all, the groups at least if not the individuals, can do their own information, breaking the power of the masters of media.

The proposal of the book is:

To denounce the authoritarian use of mass media and promote their use by the masses, with examples of alternative experiences and useful and concrete information. . . . The subject "we" of the book does not refer only to those who have participated in writing it, but above all to those whom the class at power tries every day not to let communicate one another, as they know that communications is the first step towards growth and organization. The most important media are all of us. (Faenza, 1973, loosely translated by Paolo Beneventi)

Faenza was far-left oriented and he thought above all of workers, students, alternative movements, on the long wave of '68. In fact, the technical part of that booklet was taken, re-printed and distributed by the powerful metalworkers' union, with an explicit invitation to use the video recorders as tools of counter-information in factories.

Those devices, to tell the truth, as well as very expensive if compared to those of nowadays, were quite primitive, heavy and uncomfortable, black and white tape machines bad for quality and without a recognize standard, poorly suited for editing. But it was only the beginning. Within a little time, strong standards came out of video-cassettes: *Betamax* by Sony in 1975 and *VHS* (Video Home System) by JVC in 1976 (and Video 2000, actually the best, but probably too late, by Philips, in 1980!) The video revolution was now on the market!

Act 2: Personal Computers – New Machines Meet Humans

Periodically, with a tinge of national pride, here in Italy someone remembers that the first machine working like a personal computer was the *Programma 101*, made by Olivetti and presented in New York on October 14, 1965. It was also bought in 55 pieces by the NASA and used to draw the maps of the Moon and the flight paths for the Apollo missions.

But it was still based on non-integrated transistors and the true, global change could happen only ten years later, after the invention of the *microprocessor*, available from Intel since 1971-72.

They were the hobbyists, not the companies, who had first the idea of putting a microprocessor into a box, and they were young students who dared to write a code for a device which its own creators thought was still not programmable. The Altair 8800 by Ed Roberts - no screen, no keyboard, only levers and buttons - was the device presented for the first time in the history as a "personal computer" and the young Paul Allen and Bill Gates wrote the BASIC program to make it run. It was the 1975 and the world was going to change.

In very few years, from mounting boxes PCs come as full working machines to the big market and consumers begin to get used with names like Apple, Commodore, Atari, Tandy, companies often emerging from the very artisanal work of very young people. Apple II, from 1977, not at once as many think, but over the years, is the first big commercial success, and in 1981 the giant IBM, the absolute master of computers on the planet up to then, makes its own PC, maybe reluctantly and without much conviction, but setting a standard *de facto*. *The* Personal Computer as a proper noun, if not a trade mark, is now that by IBM, thanks to the huge power of the company and the partnership with Microsoft of Gates and Allen, who "recycle" an existing operating system and make it available for any producer. As all the personal computers of the first generation had the problem to be in fact *incompatible* one to another, that was the very winning move, which made it possible to lower the costs of the various components, cards and peripherals, and to develop software that could finally run on different machines.

PCs are small, quite inexpensive and easy to use, nothing to do with the old mainframes based on punched cards and guarded by people in white coats, neither with the old stupid video terminals, which do not work at all if not connected to a network. They spread in the offices, but soon also in schools and homes, also in the form of cheap "keyboards" with colors and games, which use a normal TV screen as monitor, the so called "home computers", popular among children and families: Commodore Vic 20 (1980), Sinclair ZX Spectrum and Commodore 64 (1982), the MSX "family" by the Japanese brands (1983), and many others.

System commands during some time go on being textual, and to learn something of the often built-in BASIC language is possibly required, but software as word processors, spreadsheet, data base, as well as video games, are soon easily approachable by common people, with the only annoyance

Figure 1. Personal computer sales from 1977 to 1982. The most successful PC in the very first years was Radio Shack TRS 80, by Tandy Corporation.
(source: adapted from Reimer, n.d.)

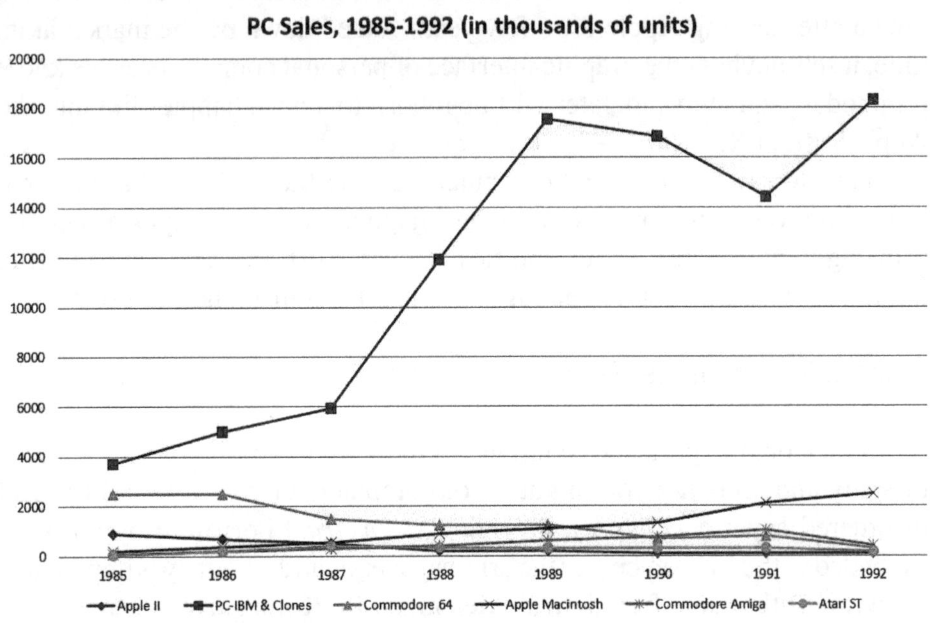

to deal with some control commands from the keyboard who one can learn by heart or read on the screen as needed (combinations of keys as Crtl + C for "copy", Crtl + V for "paste" and many others, depending of the recalled functions).

PCs become really for everyone with the introduction of the graphical interface, icons and mouse, an old idea of Xerox (the "Alto" workstation, 1973), first proposed commercially by Apple in 1983 on the too much expensive and unsuccessful *Lisa,* but getting a worldwide hit a year later, with the *Macintosh,* and implemented soon on all the other systems, in times when many different ideas are still developing, before the fatal selection that would have saved on the market, among all that group of pioneers, only two operating systems, Microsoft Windows and Apple Mac OS.

Atari ST is on the market in 1985, the first PC with a MIDI interface, showing new amazing possibilities for music. The same year, *Amiga* comes out multimedia and multitasking ten years before Windows, provides true "plug and play" connections with any peripherals and astonished graphic capabilities, which make it for some years, on the high side, the ideal machine for TV studios (in the forefront of multimedia and 3D graphics) and in the low one the best home PC for video games (with the consequent negative label of "non serious" machine). Acorn *Archimedes* in 1987 is already equipped with a RISC processor, anticipating the generations of *Power PCs* of the Nineties by IBM and Motorola. *NeXT* in 1989 is a hi-end PC, Steve Job's dream after leaving Apple for a few years. Voted down by the market at its time, it sets anyhow the graphic interface of personal computers as we know it up today, and ten years later will be reborn to a new, happier life into the Apple System X.

That's to say that personal computers, during the Eighties, have a great impact not only on work, but also on imagination and thought of people, showing a huge range of possibilities of every sort, that anyone can taste, try, develop, just from the meeting with surprising machines and software.

Needs, Possibilities, Market, Ideology

In my personal experience, coming from a classic and humanistic course of study and knowing nothing at all of computer science, after a little and disordered experience with a Sinclair *ZX Spectrum* borrowed by a friend, expanded with a true keyboard (the original one, small and soft, was impractical for writing) the same friend, a guy dealing with PC magazines, gave me an

Figure 2. PC sales from 1985 to 1992. Up to the beginning of the Nineties, other operating systems were in competition with those of Microsoft and Apple, such as Commodore Amiga and Atari ST.
(source: adapted from Reimer, n.d.)

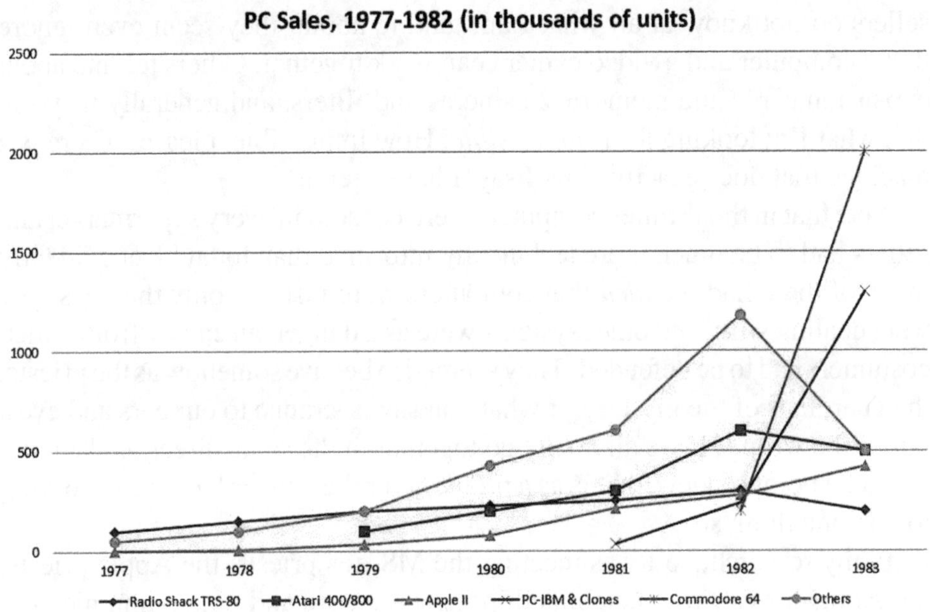

Olivetti *M10* (a Kyocera make, originally) one of the first laptops of history, less than 2 kilograms of weight and a gorgeous keyboard. Is was 1986 and, after buying a dot matrix printer, I stopped using the typewriter forever.

Two years later I got my first own computer, a *NMS 8280* by Philips (system MSX2, an originally Sanyo make), that was not the top of technology, but had some features not available on other even more advanced and expensive machines. It allowed me, only connecting normal audio and video cables, to play with video in real time, digitizing, superimposing titles and other great things almost impossible not only for MS-Dos and Apple PCs users, but also for those who managed the newer and "specialized" *Amiga*, without adding to the basic machines expensive cards or learning complicated procedures.

I began to go in schools, even in kindergarten, making children play with computers, videos, cameras…

This is an important point. It happens sometimes that possibilities meet our needs, or desires, but we cannot go on straight because of the limitations of the market or, to say batter, of the ideology which inhibits many people actually to see all those possibilities, so that the new stuff is usually framed in old, or anyway already known mental patterns.

So, in those times I'm going around in stores asking for eventual alternatives to that "video computer", whose technical limits I do recognize. But most sellers do not know at all what I am talking about, they seem even ignore that a computer and a video camera can work together. Others tell me about expansion cards and memories, cameras and filters, and generally they say that what I'm looking for is *impossible*. How impossible, please? There is a machine that does exactly what I say, I have seen it!

Note that in those times computers were not sold in every supermarket and sellers had to be much more technically informed than today! But evidently many of them had the *idea* that computers were just and only the ones they were dealing with, and other systems were as a danger, an enemy from which costumers had to be defended. They seemed to behave somehow as the priests, the guardians of the mystery: if what you say is strange to our ears and eyes, it must be wrong! If we all do not go together in the same direction, keeping the same point of view, that is as a problem for the new order we are working to, or something so.

In my research, so I was meeting the MS-dos priests, the Apple priests, the Amiga priest, who celebrated in their separate and not communicating temples, equally devoid of imagination and curiosity. I was reading meanwhile on magazines detailed articles, addressed not to media professionals, but to amateurs and schools, where special devices, very expensive, were described that would make the computer's VGA signal of PCs compatible with the composite video signal of televisions and camcorders, as an essential premise to start working on video from a computer (with ad hoc proposals of turnkey video labs, in lateral boxes: were they just ads?), and not a word about the existence – as a different, maybe not so good but much cheaper possible choice - of other kinds of computers already using a TV compatible signal. Then, it was always possible to say – it had to be said, it was a problem of those old times - that the overall quality of a MSX2 was not the same of a Macintosh or an IBM PC, and that composite video signal, if good for making titles for the shots from our analog camera, was not as good for the eyes in other uses of a computer as word processing and graphics This was useful information, so that at that point, depending on what we needed, we could make our choices.

But a sort of "fundamentalism" has always been heavily looming on the IT market, on the vendors side (my stuff is anyhow the best and all the other is nothing – and there is no wonder in this, as their aim is just to sell), but also on the users side, as giving commands to a computer is probably too much responsibility for many and to have a "big brother" who takes decisions in our place can be apparently reassuring.

Hardware and software so are to be understood and learnt and never questioned; if something goes wrong, the fault is always of the clumsy user; apparent contradictions or inconsistencies of the system are due to our inability to adapt; and with what right a common mortal can possibly doubt, with his poor observations, of that absolute perfection that are PCs and digital devices in general?

I cannot know directly, but I guess that also many readers of mine have met more than once people who, in front of a bad comment about the efficiency of a renowned software or operating system, have answered as they were personally insulted. So great companies as Microsoft or Apple have often had many not requested defense attorneys, passionate and for free, among average people.

I remember that during a book fair in Turin, many years ago, I had presented a title I had just published about the use of computers with kids (Beneventi, 1999). In it I had described activities and possibilities, and certainly not recommended of "following instructions", so much so that I had written it mainly basing on my experience on Amiga systems, which were already out of the market and had suggested, at their best times, an approach rather different than Windows (or better MS-Dos, before 1995, even in graphic dress) or Apple systems.

A teacher comes and congratulates me, having read the book and telling me that it will be very useful for her didactic activity. We begin to talk and, at a certain time, I say something about some lacks of Windows 3.1. I am sure I had not made a strong attack, but only probably quoted some of the known problems of the system of Microsoft, used as I was and I am to think that considering what is good and what is less good in any complex system, though overall well working, it is useful and important to improve it. I remember the face of that woman completely changing expression. Apparently, in front of her, there was no more the author of a book she had liked, but a guy who dared to question her absolute faith in the only way of conceiving computers she knew. And I had probably met one of those people who understand very little of computers except what they are accustomed to doing, who never ask questions and, for accepting those machines in their life, need to *believe* in them unconditionally.

Fundamentalists for interest or belief do not take into account that almost all the progresses in this field have been achieved just by people able to imagine the difference.

It's easy, then, for guys like Steve Jobs to get a reputation as a visionary, in a world peopled of aspirant bureaucrats without horizons, of women and men overwhelmed by the abundance of means and possibilities, afraid to choose and take responsibilities. But to have too much, it does not always automatically mean more freedom.

From Ivan Illich's *Energy and Equity*:

Underequipment keeps people frustrated by inefficient labor and invites the enslavement of man by man. Overindustrialization enslaves people to the tools they worship, fattens professional hierarchs on bits and on watts, and invites the translation of unequal power into huge income differentials. It imposes the same net transfers of power on the productive relations of every society, no matter what creed the managers profess, no matter what rain-dance, what penitential ritual they conduct. . . . Such a process amounts to public guardianship over a means of production to keep this means from turning into a fetish for the majority and an end for the few. ... A society that tolerates the transgression of this threshold inevitably diverts its resources from the production of means that can be shared equitably and transforms them into fuel for a sacrificial flame that victimizes the majority. (Illich, 1974)

Going back to the "video computer" I have referred to above, here are some technical data, to give maybe and idea. It was an 8 bit machine, clock 3.5 MHz (three, dot, five!) 128 K main memory + 128 K video memory (kilobytes, not mega or giga!), a video and a sound coprocessors, two 3 ½ floppy disk drives as mass memory. I am still waiting, 30 years later, for another machine or software allowing me to do those same things so simply, working with video and kids!

I remember that after launching its line of "true" PCs with the slogan "Personal Power" (great words, by the way, to be meditated, in times of "apps for everything" taking away from us any idea of "power" and responsibility) Philips in its catalogs raised the price of that "impossible" toy (almost as to say: don't buy it!) and within few years stopped the production not only of its rebadged Japanese home computers, but also of all the line of "serious" compatible PCs.

A BASIC Approach to a New Literacy

Up to the half of the Eighties, all personal computers, before Macintosh changed the cards on the table, were provided with an in-built or attached BASIC programming language.

Classical BASIC is an easy and powerful hi-level "interpreted" language (to be distinguished from "compiled" languages). Commands and instructions derive directly from English and, though it is not considered very clean and effective (someone nicknamed it "spaghetti programming") it is in fact the way across which all a generation of computer users has at least glimpsed the possibilities of what in a word is called "coding". That is to say how to give instructions to machines about what they have to do.

"Digital language" appears to be in fact very pedantic. We have to break up every action into a series of elementary ones, to pay attention to a lot of particulars that we in daily life do not care at all, and also to give a great importance to the form, commas, dots, parenthesis, positions of every word or figure, always minding the exact syntax. It is logic and math at the deepest, "basic" level, the way how digital machines *think*.

Many other programming languages exist, more or less efficient, difficult, or specialist, used by professional programmers for many different purposes, whose some names are: FORTRAN (from 1954), COBOL (from 1959), ALGOL (1958), Lisp (1958), C (1969), Pascal (1970) C++ (1980), Perl (1987), Java (1995). From the years when they begun to be developed, we can see that their story is far anterior to the appearance of personal computers. Some of them are deeply changed in the meanwhile, as the old COBOL, born in the Fifties for business use and from 2002 become "object-oriented", as well as Visual BASIC (1991), which only in latest versions supports objects, but it's anyhow something completely different from the classical BASIC (first developed in 1964), with its numbered lines, and is particularly suitable for designing graphic interfaces.

The so called hi-level languages are quite similar to human languages, whilst at the lowest level there are the *machine languages* (o *machine codes*), based on binary or hexadecimal code, primitive and hardware-dependent:

While easily understood by computers, machine languages are almost impossible for humans to use because they consist entirely of numbers. Programmers, therefore, use either a high-level programming language or an assembly language. An assembly language contains the same instructions as a machine language, but the instructions and variables have names instead of being just numbers. (Machine language, 2017)

What computers have to do can be designed also using the so called "authoring programs" popular in the Nineties (of which I write here in Chapter 3), or software like *Scratch* (from 2005), widely used today also in primary school to train kids to *coding*.

Scratch has been developed by the MIT (Massachusetts Institute of Technology) Media Lab, where in 1967 it was created LOGO, the first programming software suitable for children. LOGO was at the same time clean, easy and "structured", soon available in many national languages: a great occasion for young humans, but non only, to get used to stand in front of the "digital" world as it is, saving us from a lot of misunderstanding and weird tales, which especially during latest years have been making the relationships between humans and technology much more confused than they would be.

Giovanni Lariccia, the man who in the early Eighties brought the *Logo* language in Italy, so remembers: "The name and the theories of professor Seymour Papert, the mythical "inventor" of *LOGO*, began to be very popular in Italy around the end of the 60s. Professor Papert series of Mit LOGO Memos, in particular, were felt to be very provocative both in the educational and in the academic environment. I was enthusiastic about these theories, as well as a small group of young university professors and researchers. Then I spent one year Visiting Scientist at CRLT, the Center of Research on Learning and Teaching of the University of Michigan where the LOGO Language was compared with and contrasted to other strategies for using computers in education, such as Computer Assisted Instruction (CAI) or Computer Managed Instruction (CMI)". (Lariccia, personal communication, 2017).

These are the first lines of *Mind Storms*, by Seymour Papert:

In most contemporary educational situations where children come into contact with computers the computer is used to put children through their paces, to provide exercises of an appropriate level of difficulty, to provide feedback, and to dispense information. The computer programming the child. In the LOGO environment the relationship is reversed: The child, even at preschool ages, is in control: the child programs the computer. And in teaching the computer how to think, children embark on an exploration about how they themselves think. The experience can be heady: Thinking about thinking turns the child into an epistemologist, an experience not even shared by most adults. (Papert, 1980)

And also:

I too see the computer presence as a potent influence on the human mind. I am very much aware of the holding power of an interactive computer and of how taking the computer as a model can influence the way we think about ourselves. In fact the work on LOGO to which I have devoted much of the past ten years consists precisely of developing such forces in positive directions. . . . We have invented versions of such machines in which powerful ideas from physics or mathematics or linguistics are embedded in a way that permits the player to learn them in a natural fashion, analogous to how a child learns to speak. . . . I have invented ways to take educational advantage of the opportunities to master the art of deliberately thinking like a computer, according, for example, to the stereotype of a computer program that proceeds in a step-by-step, literal, mechanical fashion. . . . Some children's difficulties in learning formal subjects such as grammar or mathematics derive from their inability to see the point of such a style. (Papert, 1980)

Let's Not Talk Nonsense of "Digital Culture," Please!

Being not a programmer, I try sometimes to compare classic BASIC with LOGO, in order to give a little idea to nowadays public of what it is and how it can have different appearance the "digital" (that is not to push icons or scroll pictures on a touch screen!).

For example, to draw a square, with the Microsoft BASIC of that my old MSX2 (which was a powerful "dialect" in the big family of BASIC languages and used alphabet letters for complex graphical and sound instructions), I had to write:

```
10 SCREEN 2
20 LINE (50,50) - (100-100), 15, B
30 GOTO 30
```

That is to say: I choose the screen mode 2 among different possible and, inside the instruction LINE, I give a color (15) and set a "block" (B) with given coordinates. Line 30, invoking itself, is to keep the square visible on the screen.

Another way of drawing a square is:

```
10 INPUT A
20 SCREEN 8
30 DRAW "D=A;R=A;U=A;L=A;"
40 GOTO 40
```

Where "A" is a variable, the length of the side of the square and, inside the instruction DRAW (but only in that BASIC dialect) "U" stands for *up*, "D" for *down*, "R" for *right*, "L" for *left*.

If we think that some humanist in crisis in those times came to define BASIC as the "Latin of the future", we understand how the entrance of "digital" in our lives, today as well as in the Eighties, should be metabolized a little better before we venture into predictions about the possible implications in the whole of our society.

LOGO language, on the other side, did not consider the point of view of some guy looking at the screen, but always, from inside, the subjective point of view of the "turtle", which is, with its movements, the actual actor of everything we set up to happen. So, if we want to draw a square, we give instructions like these:

```
FD 100 RT 90
FD 100 RT 90
FD 100 RT 90
FD 100 RT 90
```

Or also:

```
REPEAT 4 [ FD 100 RT 90]
```

This is an example of a simple, essential "digital" thought, which anyone can easily compare with the clumsy and redundant syntax of BASIC. RT is for "turn right" and 90 can be only referred to degrees, the magnitude of a right angle; FD is for "forward" and the number is the length of the step, in this case four equal steps as four and equal are the sides of a square. And even who has never known anything of coding, at this point notices that something can be improved, because the program, in this way, draws only one square, with the fixed size of 100. We need a *variable*, which is to be given before the program runs. We call that variable "size", so that we get this program in the end, called "square":

```
to square:size
repeat 4 [fd:size rt 90]
end
```

And as the machine, in LOGO language, "learns" definitions introduced by the primitive instruction "to", next time we'll have only to type "square" and, after a "?" appears, we will type the length and have the square drawn. We can naturally also directly type "square 100", or "square 80" to have a square drawn of the chosen size.

And even who has never known anything of coding begins now to think and set possible problems, requiring consequent solutions as, for example: "Which color the square is to be drawn!" "How thick have the lines to be?" "If we do not type the side length at once, instead of a generic question mark, can we have a phrase explaining exactly what required?" Because someone could know not enough about squares, or it is not very clear if different programs, asking different questions, always use only a question mark. So, in this case, it may be more convenient to have by the program typed on the screen: "Which is the length of the side of the square?"

Anyone can notice that we are now not learning instructions about something we do not know and have to take as it is (as too many times it happens), but looking for solutions to our own requests. And we are going to like the coding language if it responds smartly and sharply, if it allow with its available commands to do what we wish, or we may think that it does not work so good or can be improved if there are not yet solutions matching our observations. The path has begun from mere consumption to conscience and, especially if we are doing this job with children, at this point they have probably started to cite icons, menus, possible choices given by the software they use and know, all enlighten by a new awareness: "It's like when…" "It happens something so in that program, in that game!" "Ah, that's how they did!"

All that is not exactly the same meaning of "digital" used referring to kids, middle aged women, grizzled pensioners all snapping with their cells on the wagon of the subway; to the new procedures we have to learn to connect with the public administration, pay taxes, enroll children in school on line; to the possibility of buying things and tickets directly from our laptop, tablet or phone. These are indeed important elements of novelty in our lives but, if we think just a little, they only allow us to go on just doing the same things we did before, using now new, probably better but sometimes annoying, almost universal tools, whose real potential probably remains unknown, if we go on to use them only to do the same things as before.

And some questions arise.

Taking in hands devices working more and more automatically, does it mean to get digital competences? Reading newspapers on a screen does it mean digital culture? Booking a medical examination from a cell does it mean digital citizenship? Or rather the exaggerated use of the word "digital" every time we touch a screen puts many people uneasy, makes them feel inadequate or old fashioned if they do not follow the stream of technological market, creates a useless contrast between the world approached through machines and the world perceived with senses.

Again, we are often driven not by our curiosity, desire, intelligence, but above all by ideology.

If Even Monopolies Have to Change Their Strategies

Probably, when they begin to play with technological gadgets, phones, cameras, interactive screens of every size and type, young children guess a huge potential power behind all that stuff and imagine possibly to be in their near future, growing and learning more, active actors and not only consumers in the information society. This is also probably what put them soon uneasy, when they discover that the "world" is not interested in general in their possible activities and that all "that stuff" is made in part to be used only according to rules always decided by others, and in part to be consumed in leisure, before being replaced in a little time by other stuff made to be used according to rules decided by others or consumed in leisure. There are few or no ways at all of understanding more, going beyond, possibly *producing* something earnest or *interesting* – as children love to say - finding out original or at least more personal ideas to break that vicious circle, to express a reality which should be not only reduced into clichés, "trends" of market, or virtual spaces where all participate and fuss and no one counts in the end.

But following rules and enjoying leisure cannot be the only aim of life, the only goal of humans; as nowadays children (as the children of ever, as the women and men of ever) aspire to feel themselves alive, playing, knowing, loving, discovering, making things and touching them, with the whole of their minds and bodies.

And when to the perception of living a life whose control is anyhow not in our hands, a general uncertainty is added about the job, the security, the future for us and our family, or for the fate of the planet itself, many personal and social problems of our times are not so difficult to be explained.

But is it all this necessary? Is it the only way the world can go?

There are big and important examples of how just in our times the possibility for average people to "change the world" is stronger than ever, examples which are carefully underestimated, hidden of forgotten, even when they concern not marginal or residual sectors of the society, but just what have changed a great part of our lives.

1. Example One: Personal Computers

It can be interesting to go first and read what was written, for example, at the voice "elaboratore elettronico" (computer), in the fourth volume of the *Enciclopedia Europea*, published in Italy in 1977. I translate:

The large-scale integration of circuits (LSI) has led to bringing together in a few millimeters card most of the typical features of a traditional central unit. These are microprocessors, which, connected to memory platelets (also made in LSI), can form miniature and low cost processing systems (microcomputer). . . . The development of various levels of processing tools (from microcomputers to mainframes, through mini-computers, small-scale machines, born around 1965 and already very widespread) can favor, together with developments in data transmission links, the tendency towards a distributed IT. Data processing networks linking private users (such as time sharing services) or university computers already exist, or are in the process of being developed; And similar solutions will allow, in a large company or organization, instead of centralizing processing and storage functions, to distribute them at various operating venues (each with their own processing tools), without sacrificing their coordination and integration. There are also those who predict that in this way we can get a general computerization of the society, and "computopolis" projects, that is computerized cities in a series of functions and services with terminals located directly in the houses, have already been taken in consideration in Japan. (Rapelli, 1977)

All in all, these visions of the "future" appear quite timid and not ambitious, compared to what is going to happen within a few years. At the end of 1976, in the world they have been installed about 250.000 mainframe computers, with IBM holding alone between 60 and 70% of the market (Capovani and Romani, 1978). Minicomputers, as the quoted Olivetti *Programma 101*, have changed something, but not the overall perspective of an expansion from the center to the periphery, according to a classic industrial model. So, though a possible wide expansion of microprocessor based devices is forecasted, it is natural to go on talking in terms of "*time sharing* services" and "*terminals* located directly in the houses", taking the words and the context of mainframes and networks.

But just in the same year of that encyclopedic voice, 1977, personal computers are getting their first great commercial achievements. And they are not only "low cost" and "distributed", but in fact fully stand alone machines, possibly not depending at all from a center. They are going within few years to spread everywhere, and from anywhere in the world people is going to have the opportunity to participate directly in the society of information, possibly sometimes as protagonists, no matter if they are young, not wealthy, far from the centers of the industrial and political power. With personal computers, center and periphery often mingle and mix and great companies have sometimes to go after the ideas of isolated and unknown students.

PCs producers are at the beginning not much more than artisans, but they conquer soon a huge sector of the market, or better, they set up from scratch a new huge sector of market, in which IBM itself after a while is almost forced to enter, however becoming soon for years, after setting its standard together with Microsoft, the best producer of PCs in the world. But – that is the point – it happens almost *against its own plans*.

An observation could be done, at this point, on a fact that is outlandish, at least. After the *LSI* cited in that encyclopedia (*Large Scale Integration*, from 100 to 10000 transistors), *VLSI* followed (*Very Large Scale integration*, from 10000 to 100000 transistors), and then *ULSI* (*Ultra Large Scale Integration,* up to 10 millions transistor and more), but today, forty years later, PCs millions times more powerful than those of that first generation come back to be considered by most as useless if not connected to a network: that's really bizarre!

But the most remarkable fact remains another, and it is worth to be stressed once more, as it is very different from what we generally expect, even from what we think possible, in times of huge and ruling industrial power. In fact, the possibility exists that something extremely strong and innovative could be born also out of the industry projects. That it is, regarding the first PCs, just what happened. In other words, distributed technology, once produced and available on the market - from the industry and commercial organization of course - it can also be used according to other purposes than those initially previewed by its makers and distributors and, in some cases, it can affect with important changes the industry and commercial organization itself.

This happens when people become aware of what they have in their hands, go beyond their designed role of mere consumers and begin to turn the means of the overindustrialization from fetishes back to useful tools for doing things again, so that knowledge and engagement can overturn the social processes and the "sacrificial flame" described by Ivan Illich.

However, until the industrial organization of the society represents the dominant system, what has come for a while out of the plans is often recovered by the big companies, in a mix of advertising, ideology and "common sense", as it happened just in this case, to the point that many sincerely believe that IBM "invented" the personal computer.

One could think that the story of PCs is a lucky case, an exception, but in the middle of the Nineties, pretty much the same happened to Microsoft, the new monopolist of the planet. They have conquered the world market thanks to a standard adopted by the great majority of the producers of personal computers, that means good and cheap machines, as well as compatible software, cards and peripherals, but the old operating system, MS-Dos, though provided with a graphic interface, Windows 3.0 and 3.1, is poor indeed if compared, for example, with Apple OS or Amiga. None of those two proprietary systems, though (which run alone) can be a real competitor to Microsoft, as Apple hardly survives (other times!) and Amiga, after collecting an incredible series of industrial and commercial mistakes, is on the way to the end.

When the finally updated operating system created to replace MS-Dos is announced and beta released with the name of *Chicago*, a competitor actually exists, possibly strong, as running on the same machines and proposed by a great brand: *OS/2* by IBM, deriving from a common project abandoned by Microsoft on the way. But the power acquired by the house of Bill Gates is so great, as well as the confidence of the users in his brand (or the fear of changing "big brother" again?) that the announcement of a new *Windows* coming as a full operating system is enough to block the possible expansion in the market of the already available OS/2. However, the release of Windows, initially scheduled for the end of 1994, delays.

In that time Bill Gates is publishing a book, *The Road Ahead*, on the "future" of computers (Gates, 1995), in which it is hard to find the word "Internet". (Staglianò, 2000) But in the universities and among the young, crowds of people every day are browsing the *World Wide Web*, creation (1990) of a team lead by a British researcher, Tim Berners-Lee, and there is a company called Netscape which is selling a lot of copies of its browsing software *Navigator* all around the world.

The Web is the very new frontier of computers, and on the web the operating system running on local machines in fact does not matter! At Microsoft, all so engaged in designing for their PCs a graphic environment based at last on a multitasking system (as the Amiga, since 1985), they have apparently underestimated all that.

They take measures, buy an existing browsing program and work hard in order to fix it and connect it to "Chicago", so that they run well together. And *Windows 95* comes out in the end "in bundle" with a browser: *Internet Explorer.*

Another time, due to the action of not only a basic, but in this case also a nonprofit movement (the world of study and research, the students of universities), the most powerful company of the planet has been forced to change its plans.

This is – and I do believe that we should think about it a little more - a very strong demonstration of what is now actually possible with the democratization of technology (hardware and software more and more powerful, cheap and easy, and more and more suitable for sharing, across a global network spread worldwide). But, maybe because, measured in human times, it is very recent, maybe because of the dominant culture of mankind is still from books and above all from TV, this possibility is quite unknown by the majority of the unaware consumers. And if we ask around - just as it happened with PCs and IBM - most of people probably think that the World Wide Web was invented by Microsoft.

Personal computers, the Web and also the operating system Linux, non profit but boldly competing with Windows above all on the market of network servers and in the industry, are only the tip of the iceberg of all a movement of open, free, non commercial or anyhow "poor" production, especially of software, without which our digital devices would be much poorer and ineffective than they actually are. From the beginning, in the Seventies, with crowds of hobbyists writing for passion every kind of programs, often bought, imitated, copied by the big companies, through all the movement of the open and free software (Williams, 2002), until today with the independent writers of very good apps given for free (of for some ads) on mobile devices, there is all a history born and grown outside and far of the plans of monopolies, and not only complementary, as able to match some niches of users maybe at the beginning out of the big market, from which sometimes great ideas and solutions for all develop and become winning.

Non profit does not mean necessary that people do not earn from their activity, as when the work is professional, not from the sale of ideas and from copyright, money is made offering other kind of services, as we will see better in chapter 6.

More in general, all around the traditionally intended industrial and centralized production, a galaxy of small makers exist, nonprofit, small profit or "quick-growing" profit (many ideas are successful in very short a time, so that also their producers possibly change field!), so that in fact "Eighty-eight percent of firms use a third-party solution provider for at least one component of their digital transformation" (Forrester Consulting, 2015).

REFERENCES

Forrester Consulting. (2015). *Digital Transformation In The Age Of The Customer*. Author.

Faenza, R. (1973). *Senza chiedere il Permesso, come rivoluzionare l'informazione*. Milano: Feltrinelli.

Illich, I. (1974). *Energy and Equity*. New York: Marion Boyars.

Illich, I. (1978). *Toward a History of Needs*. New York: Pantheon Books.

Machine Language. (2017). Retrieved on September 14 2017 from http://www.webopedia.com/TERM/M/machine_language.html

Papert, S. (1980). *Mind Storms*. New York: Basic Books.

Rapelli, G. (1977). Elaboratore Elettronico. In Enciclopedia Europea (vol. 4, pp. 415-416). Milano: Garzanti.

Reimer, J. (n.d.). *Personal Computer Market Share: 1975-2004*. Retrieved from http://www.retrocomputing.net/info/siti/total_share.html

Williams, S. (2002). *Free as in Freedom: Richard Stallman's Crusade for Free Software*. O'Reilly Media.

ADDITIONAL READING

Beneventi, P. (1999). *Come Usare il Computer con Bambini e Ragazzi.* Torino: Sonda.

Capovani, M., & Romani, F. (1978). Informatica In: Enciclopedia Europea. Milano: Garzanti.

Gates, W. H. (1995). *The Road Ahead.* New York: Viking Penguin.

Staglianò, R. (2000). *Bill Gates: Una biografia non autorizzata.* Milano: Feltrinelli.

Chapter 3
Let the Children Play!
Born Into Technology, Beyond the Transmitted Knowledge

ABSTRACT

Born in the middle of technology, kids today are "multimedia beings", naturally dealing with electronic devices in a way that is puzzling for some adults. But many experiences with children along the years can be useful to enlighten an approach to technology not only of consumption, integrated in our "normal" life, as a part of the discovery path that every baby walk, watching and touching the world around, people, nature, objects and environments, and that many adult feel the need to recover even in their mature life. Examples are given of meeting with software and hardware, lost or forgotten technological experiences, false contrasts between the new and the old, confusion and shallowness due to the overwhelming abundance of things, need maybe to go back and "play" again, in order to fix some problems between the humans and the "digital."

INTRODUCTION

Children Can Open Windows

In present society culture and knowledge are a fuzzy mix of printed books, newspapers and magazines, multipurpose digital devices and applications and above all the often forgotten big one, television! It is common to all

DOI: 10.4018/978-1-5225-3770-0.ch003

generations, but kids possibly can teach well how to *play* with all that stuff, as well as with body and nature (Beneventi, 2009), opening unexpected windows on a possible and necessary multimedia, multicultural, multicenter permanent education for everyone. In order to use properly and quickly, at a basic level, computers under any form, or video equipment, we do no need to attend to courses or special schools, as it is all already resident in the "latent culture" coming from the deep TV training in which everyone of us has grown. If we let the children free from strong consumerist suggestions play as they naturally play, looking at them we can easily "remember" that culture and use it.

The fairy tale of the "digital natives", in my opinion has hurt the kids and education in general, giving a "genetic" alibi to the inability of many teachers and of the institutions to have good relations and talk with the new generations. Much better is the definition of "multimedia beings", pointing out the cultural context where since many years babies are born and could naturally grow, if the society and the education would not split in fact their life (as well as the life of all of us) in many separate sections of consuming addiction.

Kids, today as always, need adults present and alive, not "digital" models, And even when they – the kids - can "teach" the use of some apps and devices, they cannot go on far alone, without steady "adult" reference points in their real life. If they are left to build their world without a living interaction with the adults, it would be probably a world with many lacks, "bugs" and problems, no matter if they are street children in a developing country or "digital natives" in Europe or USA.

If kids are allowed to move, with all their freedom, in an environment protected and "guaranteed" by adults who care of them and are interested in what they do, they really can teach a lot, surprisingly, not on shallow "digital" competences, but on life.

Multimedia Beings

Multimedia beings are the children (Maragliano, 1996):

Inside the home environment and other non-school spaces within which he lives and acts, the child (any child, today, in our village) does experience and knowledge of himself, of the other, of the world with an approach that is not inappropriate to define "multimedia." The child of this end of century is a multimedia being . . . because the logic that he uses to know and know himself and enter into relations with the others (also with adults, who though often respond in a dystonic way) is based on collaboration and integration of a great variety of means (telephone, radio, tape recorder, television, comic, book, sticker album, toy, etc.), inside of which the traditional hierarchies of cognition and of use no longer apply. (loosely translated by Paolo Beneventi)

From the Nineties, I have many old VHS cassettes, not only with videos, but also with recorded multimedia, made with kids in kindergarten, primary and junior high school. So I can certainly testify that five years old children can make multimedia naturally, enjoying authoring programs and many of the opportunities of creation and expression related to that kind of cultural environment.

I guess many people at this point are already raising objections: "How was it possible in the Nineties to record from computers to video? What do you mean with authoring programs? "

I can tell that in those years I was picking around hardware and software cheap (I am not rich and I never had big sponsors behind), easy and possibly good, and that we learned, amused, produced, often much ahead in front of what usually was made in schools in that time, mixing together video, audio, music, graphic, animation, as well as storytelling, body expression and theater, and exploring nature. (Beneventi, 1999; Beneventi and Conati, 2010)

We did not care much for halls full of computers all compatible and did not promote a standard instruction based only on the most popular software; we did not think that kids (and people in general) have *to learn* computers and other technological devices more than simply *begin to use* them and we did not try above all to take every situation under control. I can say that we simply let the children play. And we learned a lot from them.

Children like what is amusing, interesting, possible easy to understand if not necessarily to do; children like music, colors, movement, but above all they like to feel themselves as protagonist. But this does not mean, as many believe, that they do love to take part in prizes and competitions, because when there is no an adult pressure in that direction, when they are well satisfied of what they do, they do not ask to compete. To be protagonists, for the children, means to feel free of following their interests and curiosity and, above all, in the right time, to be taken into account by the adults. When the adults really pay attention in what the kids do – that does not mean to stay always after them, but to show care, in the end - they play well, study well, produce well, naturally cooperating and helping one another, and most of the problems of their age naturally, as if for magic, disappear.

Nowadays kids grow in a world full of easy technological devices, and they are attracted to them, but not in a different way as they are attracted to balls, puppets and dolls, toy cars and animals. I guess that if free to choose whether to run in meadows or play a video game, any child at peace with himself and the world most likely would choose the run.

But Kids Can't Change the World on Their Own

I once asked a group of five years old children:

Do you know newborn infants, who can't do anything. Within a couple of years they learn to walk, to speak a language, and no one is surprised for that. In your opinion, what is more difficult, to walk and speak a language, or to use a mobile phone or a tablet?

With no doubt, the kids answered at once:

Walking and speaking is much more difficult!

Figure 1. "The mouse! Kindergarten, 1997". (Source: video from the author's labs in schools)

So, why many people are marveled if a baby use a digital device?

Some enthusiast observers wonder about the easiness how even very young kids are familiar with "technology". They (or someone like them) watched and astonished during the Eighties looking at babies using a TV remote control, or in more recent times browsing videos on YouTube, even if they are not able to speak yet. So the enthusiasts fancy sometimes of new evolved beings, the "digital natives" who, thanks to the exposure to screens, grow automatically knowing much more of technology than the old generations.

Cats do almost the same! After playing with virtual fish swimming across the screen, trying to catch them and having seen the fish disappearing "under the water", I saw more than a cat going with its paws below the tablet, in order to lift it and find out where have the fish gone. But not all the cats, only the younger and digital ones!

A teacher tells me about her high-school students, all the time connected with their phones and making a lot of things on line but unable, if requested, to create a folder in a computer.

Sure, there are some teens and even children very good in everything "digital", but the majority does not have to know so much when picking up music or films from the web, attend social networks and make their (vertical) videos and post them on line at once, without any editing. They simply have got used to do such things, without need of all those competences which a fifty year old observer could presume, referring to his/her personal experience, from the lines of instructions of the PCs of the past up to nowadays touch screens.

The young have not gone across the same story and if, for example, they stay much time on the web, we cannot say if they are able, in case, to do a good search, or simply they know what actually the web is, because in order to attend only some sites, services or networks *to know* is not necessary.

Moreover, the enthusiasts generally do not take in account that there are also a lot of adults and even old people, who stay and "act" on Facebook all day long, as in a global fair to be filled with superficiality and gossip, hardly knowing how a computer should switch on and off!

By the way, the idea that the young are more keen and clever in using technology, it begins to be rather old and probably much less true today than only some years ago.

We all know that many of the pioneers of personal computers, in the Seventies, were young indeed, very young, about 20 years old. And for many years it has continued to be so, with above all young students playing with machines and software and discovering for themselves and the world many things new and exciting.

But in the meanwhile, for the universal laws of nature, the young of before were growing up and, staying at the PC playing video games, now many of them are in their fifties or sixties. Do we have to think that they have less control or competence than the new young, on the world and the technology which *they* have invented? That any five years old baby is a "native" in the digital world, and a grandfather as Bill Gates, for example, is an "immigrant"? It seems to be a rather antic and rough way of thinking.

Gobet remembers the teaching of Allen Newell, who: "warned against analyzing data without paying great attention to methods, and in particular against averaging data across participants using different methods" (Gobet, 2017).

The young of today can enjoy the "digital world" automatically, also without knowing anything of it. The young of yesterday – we should think of this a little more - did not have a digital world and so they made it from scratch.

A Strong, Unexpected Bridge From the Childish to the Future World

Things, in children's approach to knowledge, are not only true or false, but also "by pretending".

And *Imaginative* or *Pretend Play* is one of the main instruments that kids use to re-build what they know about the world.

Pretend play is a form of playful behavior that involves nonliteral action. Although on the surface this activity appears to be merely for fun, recent research has discovered that children's pretend play has connections to important cognitive and social skills, such as symbolic thinking, theory of mind, and counterfactual reasoning. (Weisberg, 2015)

Many educators, observing kids' pretend play when they are in the group of peers, or sometimes also alone, have noticed in it many similarities with theater. Not by chance the same verb *to play* in many languages is used also for actors, and in the Italian *animazione teatrale* this was a central topic. By playing, children often express their point of view much better than with words, or with drawings, and after the advent of home video recording, what happen in kids' pretend play has not to be always told by words, but can be communicated almost directly.

Video is not *the reality*, but it is the technical reproduction more close to the reality. It can be, of course, a way of manipulating what happens, but also a great way to tell the truth. It depends from the intentions and from how we get used to manage this incredibly powerful and universal language, which humans have at disposal only since a few years. (Beneventi, 2016)

If the educators – or the kids themselves, as they technically can - would use a bit more video communications about their pretend play, or theater, or free expression, also non specialists easily could know which great and original culture kids can express when playing, much more ahead than the performances in which they act as adults – no matter if very good or pathetic, far anyhow from the "normal" reality of the childhood - for which always along the history they have been used to get applauses, in the Elizabethan boy theater as in modern TV shows. (Beneventi, 1994)

Also – extending this kind of video telling to other activities, in which they are left free to apply the same "strategies" of knowledge or expression, such as looking for insects in the garden (Beneventi, 2012), working in group about science observations or tests, writing collective stories or making wall paintings - we could compare kids' "method" of pretending to that of scientists and artists, discovering unexpected similarities (Gopnik, 1996)

Video games are loved by the children also because they often reproduce very well or even multiply many elements of pretend play. That is a great occasion if games are added to kids' normal play, a danger if they replace it.

But the infinite possibilities of simulation related to digital environments, if well managed, can open wide worlds where kids' pretend play grows also on the technological side, meeting the creativity of older programmers, as well as the ability of educators to watch, assist, coordinate the activities, across experiences in which the players can be easily turn themselves into producers of communication and culture.

Authoring Software: A Missed Revolution?

The first *Author program* was probably *Amiga Vision*, launched in 1990, together with the *Amiga 3000,* a personal computer addressed mainly to graphic and multimedia professionals. Authoring software was probably one of the best occasions (forever lost?) to turn people's thinking and acting towards an active and aware behavior, when using digital devices.

Amiga Vision (Amiga), *Scala Multimedia* and *Infochannel* (before on Amiga and then on Windows), *Media Point* (Amiga), *ToolBook* (Windows), *Illuminatus* (Windows), *Macromedia Director* (Mac and Windows)… I remember here by heart only some titles of pieces of software whose main feature was to make people able to create programs for computers, even without knowing anything of coding, simple working on *objects* put on *pages* or along a time *score*, like in a film, and applying to those objects properties, relationships, actions, links.

Being addresses mainly to professional Info points and multimedia productions, in the gold age of optical disks, some of those programs were not exactly easy and instinctive. But I can personally witness about babies working properly with the graphic elements of very "serious" software as *Scala* and *Infochannel* were (color effects, movements and transitions) at the kindergarten and kids from the fourth grade being able to deal all alone with the full programs after less than 20 minutes of introduction and presentation.

Not "What I have to do?" (the very wrong question I heard asked a lot of times during computer courses for teachers) but whole worlds to be explored, just in the same way as kids explore and learn rules and possibilities in video games. And the results can be veritable and suitable running products, mainly multimedia but also tools and games, good to be spread and shared.

Looking at junior high school students working by themselves with *Scala*, sharing ideas and problems and suggesting one another the solutions, I personally understood, at a glance, the meaning of the word "workshop".

We can think of possible *mind storms* (Papert, 1980) among kids, supported by their teachers and connected through the planet across the World Wide Web, not only studying and learning but just *producing* information, as the multimedia beings they are, naturally using in an active way the multimedia means, at a basic but already significant level. It is a very realistic and easy scenario which until now has not been developed, encouraged, seriously considered yet, maybe because the clear distinction between producers and consumers is a too strong rooted idea in our society, despite of the availability of hardware and software which in some cases can suddenly break it.

But computers, in the imagination of the most and for the market, are not to give to anyone of us the chance and responsibility to change the world. Computers are to be used properly only by a minority of elected people and to take the most towards a "future" as in the *Jetsons*, *Star Trek*, *A Space Odyssey*!

A Simple and Natural Way to "Programming"

It is to be clearly said. In normal life we do not need to be able to program, when we use digital devices, as well as we do not need to put our hands in a combustion or electric engine when we drive a car. But a basic literacy let us know that we have to put petrol or to charge the batteries of a car, and to have software running on our PC or phone so that it can work.

Like for video, if a good editing is actually a work for professionals or advanced amateurs, but to be able to cut parts of our everyday home videos make us aware of some bases of the main language of the information era, learning how to make the digital devices "to do something" helps us to imagine some of the human work that's behind the apps and to understand something of how they work. In other words, it helps us to be aware members of the present society and not only passive consumers of gadgets.

Graphic authoring software is a strong example. It is based on objects and their properties, and it does not request a strange or "digital" way of thinking, but only a point of view, rather simple, though for many quite unusual. For kids, as well as for any person with a "flexible" mind, it is easy and probably interesting.

Most of people today do not know true authoring software, but almost all know Microsoft *Power Point* or *Libre Office Presentation*. They are made originally for a different purpose, but work more or less in the same way.

In any authoring program there are different types of objects.

Text objects can be read and written, have a font, with its size and style. They have also colors, outlines, shadows, which are not exclusive of texts, but also can be properties of graphic objects, which also have a dimension, a "depth" (dpi, important for printing, not much for multimedia jobs) and a number of colors.

They all, texts and pictures, can come in and go out from the screen in some different ways, moving, appearing and disappearing. But only texts can *scroll*, vertically or horizontally. Quite easy, isn't it?

In a multimedia page, or on a timeline, we can also put sound objects, which are not visible on the screen, but have their properties as a volume, fades in and out and, of course, are to be heard. They can be associated to other objects, texts, pictures, videos, so that, for instance, the word "Hello!" can be read on the screen and listened at the same time.

Objects have also a time we give them to do their job, and we can put every object in the order we want, one after another or all together, in front or behind, on the same or a different level, with wipes or cut transitions, linked together or separated. It is more difficult to think than to do.

In a multimedia, as generally in PC software, phone apps and web pages, there are also clickable objects, pointed to other objects or pages.

Pages themselves, on which objects lay, are to be set, so that they follow an order and a time, have transitions between one and the other, links and so on.

All this is not properly "coding", but something more basic and useful, in my opinion, for the digital, or better, multimedia literacy of all.

Education to "Digital" or to Multimedia and Multicultural?

The word "digital" to say the truth, here is often in quotes because, in the end, it does not mean something exact and univocal. It should refer to the internal way of working of the machines of the present time (discrete information, depth in bit, state 0 or 1, Yes or No), but only good programmers have a true idea of it. In current language people use the word to indicate recent devices, touch screens, the web, the social networks, electronic books and many other, often very different things, in contrast with paper books, traditional manufactured products, nature and senses. The real contrary of the word "digital", in fact should be "analog" (continues information, less or more of something, no division in bit but a unique flow of shapes, colors, sounds).

And our world is based not on a *digital*, but on a *multimedia*, as well as, with the globalization, *multicultural* system of knowledge and communication, more and more. That system, of course, in the present time generally relies on a network of digital instruments and codes, but it is not on the instruments that we have to focus our attention:

> *In a social environment characterized by the intertwining of different cultures, comparison and cooperation become the primary requirements. Common objectives to be sought together, as an antidote to the destructive struggle, in order to create a society in which everyone can achieve their full potential in a tolerant intercultural atmosphere. (Bruno, 2013)*

We all know the trouble for every single woman and man in everyday life to deal with a overwhelming amount of pictures, words, music, coming from TV, radio, the Web, papers, street ads. The world is apparently running at a speed very much higher than that a normal human being can stand and the crowd of many and different media leave us often in confusion. We can see that, in the end, our society does not offer an adequate literacy about multimedia and multi-culture and that the "digital" era, though providing "an app for everything", seems in fact to create new and unsolvable problems. Is it necessary?

Confusion With Everyday Tools, When We Do Not Believe Our Eyes

Problems and confusion are often also in very "familiar" situations, which people generally think that there is not much to learn more about.

We have already seen and we will see again about people who take videos with phones vertically. But also the "old" habit of watching television puts us sometimes in strange situations, related to the different possible formats of video communication.

At the present time, while I am writing, year 2017, all television sets in stores have a 16:9 wide screen, HD or 4K (that is the definition of the pictures, high or ultra high). Some televisions maybe disclosing the future (who knows?) in the ultra wide 21:9 (7:3) format begin to appear, but many people in the world probably still have in their houses old TV sets in the "square" 4:3 format (the once so called "standard definition"), that has been the format of all TV programs for decades and was abandoned only a few years ago.

Movies originally made for cinema, depending on the year of production, on the intentions of makers and so on, are in different formats, mainly 7:3 but sometimes even wider, or more frequently narrower.

All this, TV and cinema productions coming from many different directions in time and space, whether broadcasted on the air, cable or satellite TV, read from a DVD or a Blu-Ray or a pen drive, or watched directly on a web channel, hardly always completely fit in a screen. We will have many times a full screen with crop images inside, or full images leaving vertical or horizontal black bands on the screen. Some functions are provided by the TV sets producers in order to automatically recognize the format, but not always they work.

It seems rather complicate and difficult to understand, but when watching any video or TV program, we do not have to understand and know anything in fact, but only to *believe to our eyes*. If people on the screen appear too much fat or slim, if cars are flattened vertically or horizontally, if soccer balls appear oval as the American football ones, it means that the format is wrong, there has not been a right automatic correction and we probably should do something, as pressing a button on the remote control to match the exact *aspect ratio* of that program.

The problem is that many times people simply do not notice the mistaken format, or do not care of that. They keep on watching stocky and short top models and basket players, warped panoramas, vehicles and faces, but they do not feel the need of doing something to fix the error. As if more than watching a film, a variety, a documentary, a sport game, they were watching just their television set, or computer or tablet screen, perfect technological objects which are never to be corrected. We have spent much money to get them, so broad and bright, with high or ultra high definition, and it does not matter if the pictures we see inside are warped.

"The medium is the message!" wrote Marshall McLuhan (1964), in his *Understanding Media: The Extensions of Man*, published in 1964.

This affects, in my opinion dangerously, all the way how many people face several aspects of their life, as the medium, the social situation, the habit to follow a general conformism, appear to disconnect more and more perceptions and thoughts from the senses. In other words, we do not believe to our eyes anymore, as if the truth were on the screen. And that is particularly dangerous, when screens of every size multiply more and more, all around us, in every moment of our life!

I remember, after a whole movie projected at the cinema in a wrong format, in front of my questions about why so evident a problem had been not fixed, other spectators, who had not noticed anything, said to me phrases like these: "But you are an expert!" "Maybe it was a director's artistic choice!"

Toward a Flat and Empty World? Maybe Only a Matter of Points of View!

The Italian economist Luigino Bruni, in a brief and passionate book, complains the emotional weakness of present societies, due to the "almost sudden disappearance of a millenary richness of cultivation and education of emotions.

A people whose babies have been grown by TV and video games, where the biggest emotions are those of four heads monsters, or of the trivial stories of soap operas for teen agers, soon it becomes a people without a civil and democratic soul, and can awake someday inside a meritocratic technocracy which rules a world without democracy. (Bruni, 2017, loosely translated by Paolo Beneventi)

In my opinion, the danger is real, and I agree also with Bruni's idea that literature and art should be studied and practiced more and not less in nowadays schools, above all when the main trend is often of replacing them with an ambiguous and uncertain education to "technology". But if we think of the old times, horrible monsters were produced and followed also in societies whose citizens were much less "distracted" by the market; and if we think of ours, through and over the global homologating plans of some, something do happens here and there, that take people elsewhere imagining and dreaming, in not different a way from the better tradition. It can happen in hidden corners, out of the big trends of the market, but also involve the big public, and this is true especially for children.

Overall quality of books for kids is not worse than some decades ago, and rather - from what we can see in fairs and bookshops, text and above all pictures – it has generally improved. And not all toys are related only to trendy TV series or video games, as we could believe at a first, swallow glance. For what concerns storytelling, in books and multimedia, we can discuss if Harry Potter stand at the same level of Pinocchio or Peter Pan, if the Pokemons are above all commercial gadgets or something more, but kids love them and dream with them, as well as with many other characters and worlds which are not so flat and trite as many think.

The problem today for children, often not allowing them to grow in harmony, in advanced countries is rather more that they have no time and occasion to play free with the other children; and in other countries, that they lack what they need simply to live in health and peace. What they watch in TV and also their relation with technology – if we look just a little behind the appearance - is after all a secondary problem, as a secondary problem could be for every people the confusion and promiscuity in which the market put everything together in a context of mere consumption, regardless of the quality, if we only had the cultural tools to select and choose.

But there is another bigger and new problem. More than the bad level of what we consume (music, TV and cinema, electronic or paper books, web sites, e-games and apps) - as it is true that the most are probably bad, but there are also good and very good too, quite easily approachable – a new underestimated problem is the worse and worse quality of what in fact we all *produce*. We generally are not aware yet – human minds are slow in changing their mental patterns - but we all have been taken within few years into the new role of producers of information, billions of texts and pictures spread by common people all over the world on personal and social media, made mainly by women, men and kids without any literacy about the media and the languages they use, and this is affecting the overall quality of the information. Even professional broadcasting companies do not care anymore of wrong formats and warped images when mixing for example videos from different periods, or taken with phones vertically on the field. And, following the wave of average people, always posting and tweeting and anyhow looking at their cells, even politicians, chairmen and ministers, more than in the official sites, meetings, cabinets or parliaments, often reveal their ideas and actions directly on social networks.

In my opinion and experience, this state of things is though due mainly to a general but not so deep confusion, as when people have the occasion of changing their point of view – it actually happens more locally and in small groups but, using otherwise the same social networks, we could maybe turn it into a global movement – it is not so difficult to recover the pieces of our scattered knowledge and ability of being active actors in the world, more than only a jangling and impotent audience.

The first step, quite easy technically but very difficult for many, should be to find out a way to put together the books which we have read and studied, the television we have watched and the electronic devices we manage, without seeing all the contrasts between the different media, or between nature and technology, the old and the new, the grown up and the young, that often make so difficult if not impossible to move serenely in the world where we live.

Observing from outside, not living the situation and really doing the thing, not having solved our personal problems with books, television and all the crowd of the digital gadgets, every attempt of explaining anything is easily clumsy, ideological and wrong.

Another time, a help can come from the kids who, not so caught yet completely into the one dimensional consuming world, more than the adults can deal naturally with the different media, but also with their bodies, the nature, the things and living being of the world, without making too many questions, but simply as they please and they can. We have, of course, to give the occasion to kids of helping us.

A Video Camera in the Hand (the Magic Mirror and the Literacy)

It is not the same, no, it isn't!

Going the first time working with a group of 2nd grade kids, some of them with physical and mental problems, I am there with a video camera, a compact one but rather big, with a large lens, on a tripod.

It is true that I could have used a mobile phone for taking almost the same pictures, or a reflex photo camera to take possibly something better. But – let me tell this – phones are too much familiar to the children and reflex are too complicate to be managed well at once. With a video camera kids have only to mind the display and the zoom lever to set up the initial frame and… they began to imagine, to play.

"Let's make *Titanic*!" one says, and another goes and take the emergency plan of the school and shows it to the lens. Then a group begins to play with the handle of the tripod, with the camera on it, especially a long haired boy who – they tell me – always keeps on touching his face because of a nervous tic.

Many things then happen. I give the camera to the teachers, also removed from the tripod, so that they can shoot the activities programmed for that day.

But the long haired kid many times goes back to the tripod and uses it again as if to make a film. Never we have seen him, that day, touching his face. The other one, with big problems in relationships, having completely refused to work with the others, has gone to the big wall blackboard and now is drawing with a chalk a huge ship: "It's the Titanic – he says – and I have also drawn the great line of the break on its side".

Everything is documented in video, with my camera and also with phones.

But a video camera, for its "nature", is not only a machine for making videos, probably still the best for learning. In kids' imagination - for what we have seen in this and in many other occasions - it has probably a strong symbolic meaning, which makes it very *good for play*, too, for pretending to be a cameraman, a director. That's what with a mobile phone for instance, though good for actual perfect shots, does not happen. Many kids, before shooting with a video camera, without any suggestion from the adults, go and clap with hands in front of it and say "action!". They are *playing* cinema!

Symbols, related to the objects of technology, as well as to other objects, people, plants and animals, places and landscapes of our lives, they are important and often unpredictable, so personal and particular they are. And the video technology - as well as good to send instantaneous videos on YouTube – is once more a strong tool for *metaknowledge* (Evans and Foster 2011). That is to say to understand better what we do and we are.

With another group of kids, older, junior high school, after having always used for taking our videos a good full HD compact video camera, almost new, with manual controls, and sometimes mobile phones and a *GoPro*, I brought one day another camera, ten years old, standard definition, one of those with mini DV cassettes, which allowed us infrared filming, in the dark. Great for horror stories!

Many old cameras, around 2005, a bit earlier and a bit later, had very interesting digital effects inbuilt and, when it is possible, I still make the kids play with them.

Absolutely great is the possibility of mirroring themselves in real time into a TV screen, connected to the camera, while warped into a ball, a cube, a symmetric vision, or moving as ghosts or zombies or even as the super hero *Flash*, because of slow shutter, digital delay, or very fast acceleration. These are all things that I had in my old cameras and which give "in real time", to the excited children, many ideas about stories to tell.

I have a video where 6 year old children, perfectly according to the their deformed image into a globe in the electronic mirror, put their faces exactly before the lens, so that they can move and say "hallo" and "meow" and do strange voices, laughing and crying and amusing so much, but never loosing the "sense of framing": easy, instant, perfect.

I have kids "dancing" in front of the weird effects of symmetry on their faces and bodies, moving perfectly their arms, legs, heads, according to a music that we have added to that sequence *after*!

Such effects, put once in devices but usually unknown by the majority of users, are since years by many considered useless, as they could be easily applied in a second time, with a computer in post production. False! We need to watch ourselves "live" in a TV screen, we cannot mirror and adjust our actions according to the effects in post production. Children can't!

A reflection perhaps should be done, about this and other opportunities which in educational work have almost never been well considered, and which we risk to lose forever, though probably it would be enough to ask for them again to producers, at least for devices aimed to an educational use. As also for video, such tools should be proposed by educators, after testing them well with students and see what goes and what not, overcoming the usual passive attitude in front of the established "professionalism" of technicians and the unlimited faith (or complete disinterestedness) in what new technologies could actually bring in the everyday life of a school context.

Also because nowadays devices mainly used for making video, mobile phones – as we will see better in a chapter ahead – do not give the possibility not only of mirroring, but not even of watching our shots into the bigger screen of a TV set.

In general - we can summarize - school has never really educated to television, even when it had become in the whole of society the main "teacher" of the kids, also very young, Then, it has never really educated to active video, though many very good experience have been attempted in every grade, from primary to high-school, since the Seventies of last century, which have not much influenced the institution and its official programs.

If school had thought of television and video in terms of basic literacy, we would probably have now some tools for watching TV and making video somehow special for education, provided for example with the functions I have quoted above, or others; and probably we would not have people anymore who form their knowledge of the world, their moral and behaviors, their political ideas, passively watching television.

Personally, from my direct and quite wide experience of kids playing with cameras, I can witness, having tested it so many times, that when they are not left alone with their television models, but are requested to look for quality and try to do things well, if they see the interest of the adults for what they do and they are – as it is for any activity that kids perform in school or in a supervised group – easily they take the best from their latent TV culture and are able soon to organize it, with very good results. The teacher, or educator - or whatever we want to call the adult of reference – is to be skilled enough in making videos, but above all has to propose to the kids simple things, but somehow "perfect" and surprising. It is enough, for example, instead of the usual long hand-held shootings with a cell phone, all moved and with no stops, to begin trying short shots and keeping the camera well steady on a tripod and then to watch the results. The adult gives suggestions, analyzes with the kids what is good and what is not, amuses with them, and then all together, with pleasure, without anxiety, try to do their best and shoot and edit videos not much worse than the ones the kids are used to watch in TV. Nowadays means make all this possible and quite easy (if one does not immediately tries to go overboard) and someone has to show and possibly point up it, so that the kids get aware of such an opportunity.

Figure 2. A young girl looks at her kindergarten mates through the lens of a video camera
Source: Original analog picture from a workshop in a school, Brescia, Italy 1987.

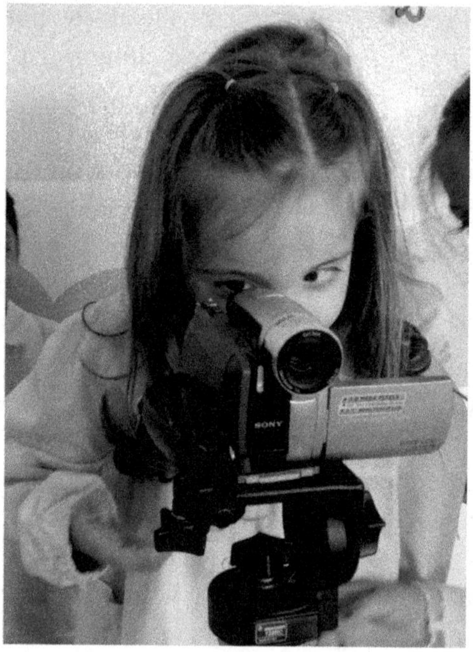

While playing, girls and boys generally give original responses, teaching also something new (useful variations on the theme, but also quite unexpected points of view) to the teacher, and begin to think themselves, more than as actors or directors - which they are not and have not to think to be - as *producers* of something, not only stories or fiction, but possibly also *information*.

What We All Could Have Discovered Simply Playing With a Computer

I have written above of the season of multimedia and authoring programs but, twenty years after many exciting experiences from kindergarten to high school, personally now I am back to make above all videos, giving up the idea of producing interactive works in education contexts. It is too difficult a task, and not for the kids, of course, but for most of the teachers and the "system".

I am sure of this: giving simply the possibility of "playing" with a PC, every kid, or better groups of kids, can learn in a few days much more than what they usually do during years of instruction at school. This happens if they meet the right software, are free of exploring and connecting their trials with the stories they like to listen and to tell, taste the possibility of obtaining real effects with graphic, video, sound, like those they are accustomed to in TV and video games. It is not necessary for them to get very skilled, it's enough to glimpse, and imagine, setting connection between the different areas of their experience and dreams.

I like to tell a scene, in a school computer room, some years ago in Italy, where I had taken the CD ROM of a "true" painting program, suitable also for five years old babies but very powerful, with many graphic possibilities which only a computer can give (as for drawing lines and filling squares with a color, people do not actually need to use a computer!). So, instant warp functions, shades, symmetries, images or part of them usable as a brush, for hand drawing or to spray them all across the screen, and so on. A teacher is trying the auto executable software, while the pupils of a fourth grade class are leaving the room. A kid looks at the screen just a few seconds and says in wonder: "Ah, it is so that they do!"

Some minutes later, a fifth grade class comes in the room. A kid looks at the screen just a few seconds and says in wonder: "Ah, it is so that they do!"

I do not know what those kids exactly did, in that computer room. Sure they were used to Microsoft *Paint*, probably since years. But evidently using Paint for years had not given to them the idea of a possible link between what they did with computers in schools and the graphic they knew on computers and video games in their "real life", at home and with friends. "Ah, it is so that they do!"

It is something like to play the recorder in schools. All kids do it, but their idea of what *music* is, out of the classrooms, is surely another!

So, few seconds can be enough for changing a point of view, for opening a mind to the world of possibilities which in this case computer graphic offers. Just as to stay properly behind a camera (but one have to know that, to experience that!) for beginning at once to make a true film; or to play for instance the first time the right notes on a bass guitar with other accompanying instruments and *listen* that they go well together!

We can do it!

Then, we are not obliged to do, we are not all so interested and so clever to aim to become all computer, video, musical geniuses, But we know now that we can do our part, we have tasted the magic of the information society whose basic tools are machines like the ones we know and we practice everyday, and whose processing works we recognize in cartoons, TV series, video games, as the high level end of what, at a basic level, is part of our living experience.

Showing only some little examples from sound software, watching together on the screen the "drawing" of our own voices, listening to them reversed, very fast, slowing down or robotics, what five years kids ask then to do is just exactly what the sound processing app allows to do. They know it as for instinct. And also the descriptions of the effects we play with are very effective, strong: "He seems to speak with an iron voice!" "She is a girl with many heads!" That is a very childish description of the effect called "choir"!

Browsing and knowing what to find, almost all the creative software known in the past, or something similar, is available also today, for PCs, tablets, phones, and creative people can actually use it and do many things, including playing with more recent technological possibilities, as virtual and augmented (artificial) reality.

But generally it happens as by chance, or following commercial trends, or always repeating after all the usual attempts of transforming something traditional (the book, the TV documentary) in something somehow absolutely new.

Did you know that an Egyptian temple can be recreated in a 3D environment and that it is possible to walk through it as a true surround environment without leaving the classroom? Or view on your screen a complete Roman house, with all its rooms, and move its figure to see all its views. (Lafuente López, 2016)

Yes, we do know! They are more than 20 years, from the first introduction of 3D graphics, that you are trying to create and propose to us something like this, in more and more wonderful ways, and I always go on thinking that, after the first amazing enthusiastic approach, it is in the end a boring experience!

Basing in practice still mainly on the everyday school tradition (more than on the ideas of the great educators!) educational systems do not seem to be able to put themselves as a steady reference point to help the young actually to link, put together and somehow in order the multiple and casual experience they can have in their "normal" life of digital worlds and possibilities, though academic studies and researches, as well as advanced experiences on the field, do give good reports and studies on it.

With technology driving 21st century education, the infusion of appropriate pedagogical strategies supported by research-based theories will enhance digital delivery, promote positive learning outcomes, promote self-reflection and self-assessment, engage all learners in the process, and provide powerful learning experiences. Courses with online learning, fully online or hybrid/blended tend to produce stronger student learning outcomes than completely face-to-face instruction (U.S. Department of Education, 2010). . . . Collaboration is a key pedagogical strategy to support effective delivery of digital learning. When promoted effectively, collaboration brings together students with different innate intelligences to work on projects as participants, creators, and assessors. This concept is expounded further in this chapter in the discussion of Gardner's (1983) multiple intelligences. (Bull & Patterson, 2017)

At School with Tetris is an evocative title, just suggesting (but why not, why so little in so many years?) a natural possible meeting between school and video games:

The computer opens for the first time the road to integrate perception and action. . . . Computer becomes a powerful metacognitive machine that allows the child to reflect and know their mental processes and strategies that are most appropriate to each one. (Slomka, 2009, loosely translated by Paolo Beneventi)

But is it still very rare that in school kids play "changing the faces", with a few movements of the mouse, as a group of kids once did, enjoying with "their" Da Vinci's Gioconda having just eaten a lemon!

Back to Book and Paper, Once Again?

A lot of words have been written during the years, about the possible next end of the book of paper, sometimes bringing intelligent arguments, but also often building real monuments to the nonsense.

It is not only because of the digital publishing and the e-books, as already at the end of the Seventies of last century, I remember that some people was already talking of the "death" of the book, killed by video cassettes! I wondered at those times about what kind of reasoning could be behind such an idea, but now, so many years later, all this can be useful for highlighting one simple fact.

From that time, video cassettes are over, musical cassettes are over, vinyl disks are somehow reborn for fans but they not exactly vital and other media which in the Seventies had not appeared yet have already ended their cycle, as floppy disks, or are only maintained for a niche use, as CD and DVD ROM.

Figure 3. Primary school kinds were playing with a 'goo' software and then 'reading' the picture and telling the story of the Gioconda and the lemon!
Source: Image processing in author's lab during "Amico Libro" fair, Castle of Belgioioso (Pavia, Italy), 1999.

Also all the devices and peripherals for computers with parallel, SCSI and firewire interfaces have had their day. Audio CDs, video DVDs and Blu-Ray discs are still on the market (I am a fan of them, as I am writing here and there in this text), but the only medium steadily alive from the Seventies till now (or better, since the Sixteenth Century!) is printed book.

The secret, and the reason why, in the end, it probably will never die, is in its unbeatable *interface*. No electricity, no cables, no operating system which can become obsolete - think only, for instance, of the Encyclopedias running on Windows 3.1, MS DOS, old MacOS and 16 bit compatible systems in general, become completely unreadable after only a few years - but the unique feature of connecting directly to human eyes.

In the future, it is reasonable that we will have a less number of printed titles than today, because print is expensive, some instant or poor publications are not worth of wasting a precious material such as paper (with a great relief for forests), and digital publishing is available to all (as maybe not many people are aware yet that everyone today can already be a publisher for himself, for free, with the possibility of reaching virtually all the readers of the world, across the libraries on line). But it is hard to think that mankind is going to give up publishing a number of chosen books on paper, probably of a better quality than the usual now, made to last in years and maybe centuries, beyond technology storms and temporary operating systems.

In such a case, it would be a very poor mankind.

By the way, news cyclically appear about advanced school systems where italic writing is no more taught, as it would have become useless by now, and the same need of handwriting is questioned, as everyone uses smartphones and keyboard for communication.

So, no more hand, pen and paper, only screens! And we already expect now to see iPads hanging from a pin on the door of the offices, showing the writing "Be right back!"

It's seems incredible that what is the result of hundred thousands years of evolution of the human species, handwriting, which implies the activation of specific areas of the brain to develop the necessary motor skills (James and Engelhardt, 2012), could be treated now by some in terms of useful or useless, only referring to everyday (and supposed: are we really sure of it?) habits of communication. It is true that among new generation italic handwriting in recent years is not much loved but, apart a question of "democracy", for which that is not a good reason for denying the possibility to a minority to use it, if they like, studies have demonstrated that for writing in capital letters or italics we actually use different areas of our brain. (Tucker, 2015). That is to say: we do not have idea of what could happen if, for a political and bureaucratic choice, this ability of handwriting in italic characters, always owned by cultured classes and conquered by almost all humans only since few years, with the spread of the mass school, were removed again. All present society, digital technology included, has been created by people somehow trained to use also those areas of the brain related to italic handwriting. Are those areas only for writing? Is it prudent to inhibit them from working?

Least because, if we look at the strong and "winning" Asian societies, every child in Japan and China is trained not only to bland writing exercises as in western countries, but to draw very complicate ideograms using a brush. And, in a school time full of everything, in which what makes it often hard to be stood by a number of students is the sense of inutility of too many activities, is it just on very basic competences as handwriting that we think of making cuts?

By the way, the idea of young generations doing everything with digital gadgets is quite a fanciful tale. Teen agers do use much mobile phones, to play games, listen to music, be connected with friends and some pages and channels on line but, when not communicating with SMS or Whatsapp, often they prefer handwriting on sheets of paper rather than to use text processors on mobile devices.

Middle aged people use word processors and other productivity software on mobile devices, as well as phone alarms, and often, looking inattentively at their sons and nephews with phones and tablets always in their hands, they are tempted to project onto them their own attitudes.

But teen agers do not awake in the morning with a phone alarm: they ask their parents to be called!

When the "Digital" Multiplies Human Errors and Bureaucracy

I am in Valencia, Spain, March 2016, in the cheap "robotized" hotel provided by the organization of a festival. I see that my room is booked for a day less, and the gentle woman at the reception says that I have to book again and pay by myself, as it is impossible for her to correct the previous information: the system is centralized and does not allow doing it! Then she says something as: "This is the future, we have to get accustomed to it! "

"No, this is not the *future*, madam, it is just a bug in the software, as some human has not thought well about all possibilities and their combinations! "

The big misunderstanding, imagining a world managed by the machines, is to forget that behind them - till now and in the next future, at least - always there are human programmers. Artificial intelligence cannot pass over a starting human point of view, and all information, if we look back, is from human sources.

It can be *technically possible* to have factories without workers, cars without drivers, papers without journalists, armies without soldiers, but it does not mean that it is necessary or inescapable. Some people do their job thinking how computers, robots and software could replace humans fully or in part in a lot of tasks, but in the end there is someone (else) who has to decide, and they will be always humans. Even if they decide not to decide, and to give all responsibilities to machines. And this is what, in a broad sense, we call *politics*.

Not a great level of literacy is necessary to know that all what is inside machines, it has been put there, at a certain point, by humans. It can be not totally clear in the imaginary world of Star Trek, the Jetsons and HAL, but it should be sure without any doubt in the real world of personal computers, where any citizen, from childhood, has the possibility of interacting with machines, to test their basic functions, to put and take out things (data) after having processed them according to his or her needs.

As computers (in their various forms) and networks are everywhere, as they are quite easy to use and to understand, as to know the alphabet of how machines work is probably a vital condition not to be dominated by the "machines" themselves, that's to say, at this stage of the technology at least and for a long time more, by a small groups of humans in whose hands the control of the machines actually is.

I have asked to professor Lariccia (2017):

Nowadays, digital devices are a significant part of the daily life of all generations, very easy, powerful and often suitable also for illiterate people. But screens are not the reality, not the tangible one at least and maybe to know what is beyond can be useful for a more conscious use of them. Is "coding", recently in vogue in schools, a possible answer?

He replied:

Not in my opinion, also because the graphic interface very often has been taken as the core business, whereas it is only an interface. Little attention has been paid to the development of the deep thought process. what A.Newell and H.A.Simon defined the protocol analysis. This kind of activity, namely recording and "decoding" the thought process would be much easier today thanks to the progress of the video recording technologies. It should be reminded that great advances in the discovery in the so called innate mathematics have been possible thanks to techniques of eye movement recording. (Lariccia, personal communication, 2017)

I am on a local electric train in Milan hinterland, Italy. All stations are new and designed according to the same, stylish way, but the names on platforms are not well lighted in the night. Probably the designer does not use to take the train himself with dark and fog. But a beautiful female sampled voice gently announces every station, whose name is also written on the displays inside the wagon.

At a certain time, something probably goes wrong on the power line and the train stops in the obscurity and silence for some minutes. When it moves again, the gentle female voice restarts announcing next stations, but from the beginning of the trip.

Digital equipment is not intelligent, it simply follows instructions, and in this case it "knows" the names of all the stations, from the first to the last and vice versa. According to the direction, it says those names in an order to be followed, but nobody has told the software to look also for the real position of the train, checking for example the data of its list with those of a GPS transmitter.

The key for creating good software is to ask the right questions, to forecast any possible variable, including in this case the possibility of a temporary electric black out and also the situation in which the names of the stations, for whatever reason, are difficult to read, so that the software is requested to be not only accessory but really helpful. Something more, possibly, than the redundant role of many Power Point slide shows, that simply repeat or resume the words of the speaker, as a mere embellishment, a splash of technology on the usual report, thought and lead just as in pre-digital times.

More than problem solving, the core point is *problem setting*. With the final result, if we give information, that it be pertinent and exact.

Tell me if it is not ideology!

Searching on line for the words "bureaucracy" and "digital", almost all the references give the two terms as opposite. That is to say, it's a common idea: digital can help us to solve the problems given by bureaucracy! Yes indeed, it can…

But what does it happen when people a in bureaucratic position, with a bureaucratic mind put between them and us web portals, digital identities, registration codes, digital forms to fill?

There seems to be few press or academic literature about the new impenetrable bureaucracy given just by a bureaucratic use of digital means and networks. But from many examples, we can guess that it is going to be in the near future a problem to often deal with.

REFERENCES

Bruni, L. (2017). *La Felicità è troppo poco, Note a margine del nostro capitalismo*. Ospedaletto – Pisa: Pacini editore.

Bruno, R. T. (2013). Storytelling in Intercultural Education. In *Didactic Strategies and Technologies for Education: Incorporating Advancements*. Hershey, PA: IGI Global. doi:10.4018/978-1-4666-2122-0.ch030

Bull, P. H., & Patterson, G. C. (2017). Strategies to Promote Pedagogical Knowledge Interplay with Technology. In *Flipped Instruction: Breakthroughs in Research and Practice*. Hershey, PA: IGI Global. doi:10.4018/978-1-5225-1803-7.ch003

Gobet, F. (2017). Allen Newell's Program of Research: The Video-Game Test. Topics, 9(2).

Lafuente López, L. (2016). *Virtual Reality And Augmented Reality In Education*. eLearning Industry.

Maragliano, R. (1996). *Esseri Multimediali, Immagini del bambino di fine millennio*. Firenze: La Nuova Italia.

Slomka, E. (2009). *A Scuola con il Tetris*. Torino: Book-jay.

Weisberg, D. (2015). Pretend play. *Wiley Interdisciplinary Reviews: Cognitive Science*, *6*(3), 3. doi:10.1002/wcs.1341 PMID:26263228

ADDITIONAL READING

Beneventi P. (1994). *Introduzione alla Storia del Teatro-ragazzi*. Firenze: Ponte alle Grazie.

Beneventi, P. (1999). *Come Usare il Computer con Bambini e Ragazzi*. Torino: Sonda.

Beneventi, P. (2009). *I Bambini e l'Ambiente: Per una Ecologia dell'Educazione*. Casale Monferrato: Sonda.

Evans, J. A., & Foster, J. G. (2011). Metaknowledge. *Science*, *331*(6018), 721–725. doi:10.1126/science.1201765 PMID:21311014

Gopnik, A. (1996). The Scientist as Child. *Philosophy of Science*, *63*(4), 485–514. doi:10.1086/289970

McLuhan, M. (1964). *Understanding Media: The Extensions of Man*. New York: McGraw-Hill.

Papert, S. (1980). *Mind Storms*. New York: Basic Books.

Chapter 4
Technology Is Not a One-Way Road:
At a Crossroads – A Real Revolution or Only Market?

ABSTRACT

Surrounded by new pieces of technology every day, nowadays citizens often do not have so updated schemes of thinking as the hardware and software they deal with. So, they go on mainly with old ideas and are not able to imagine the future beyond the suggestions of the market. Velocity and competition myths are not born exactly with the digital age, and nonstop connection could be probably something more than only to chat with friends, watching videos and listen to music on line. From different sides, in our experience often mixed into an indistinct set, commercial social networks and non-profit ventures have in fact changed a great part of our lives and habits. But the way of technology is not already drawn, as it depends much on our deeds and choices.

INTRODUCTION

Velocity and Confusion

Are we really living, in our times, a technological revolution?

Since many years many are repeating as a mantra that technology is completely changing our lives. Is it true? It depends on the points of view and

DOI: 10.4018/978-1-5225-3770-0.ch004

also on which technology we are talking of. Automobiles, television, mobile phones have surely changed our life. Personal computers would have changed it much more than it actually happened, and maybe just in order to avoid too many changes, the producers have provided other devices that people can use for many purposes instead of PCs, because in the end the humans beings do not like to change too much.

Two billions people are on social networks, but their "political" weigh is very light if not inexistent; strong examples of shared innovative productions are on line, together with a never seen before massive waste of time. A big struggle is probably happening in the world web between creativity and conformism, drawing different ideas of future.

Velocity is a very early 20th Century idea, the myth of Futurism. And about the present, referring above all to leisure time, sport in TV and web attending, some authors speak of "futurism of the instant" (Virilio, 2010; Redhead, 2016).

In my opinion, the conviction that the humans must always run, if we think well of it, is very curious, as today there are machines which can run for us: flights and hi-speed trains in the real space, data of every sort in the virtual one across the global network, while we are staying comfy on our armchairs, sleeping, or going for a walk, and in the meanwhile producing much more than with the hardest work until not many years ago.

In my personal experience, just the possibility we have today to stop when we need is a great chance for many activities, as we can rest when we are too much tired and confused and restart in the right time, so producing much more and much better.

Against the necessity of always applying "Clarke's law of continuous acceleration" to every sector of the economy and also to more and more parts of our life, free and leisure time included – as for many people the "future" requires - there are not only some romantic nostalgic for the old days, but also people at the forefront of the most important technological progresses of the present time, such as eminent protagonists of the hacker movement, as the Finnish Pekka Himanem.

He writes:

The time-optimization method developed by Frederick Winslow Taylor for industrial capitalism is still alive in the network society. From the typical information professional of our time, this culture of speed demands an ever more effective use of his or her working hours. The workday is chopped up into a series of fast appointments, and he or she has to hurry from one to the next. Constantly trying to survive some project's deadline, the professional has no time left for playfulness and must optimize his or her time in order to stay on top of it all. . . . The culture of work-time supervision is a culture that regards grown-up persons as too immature to be in charge of their lives. (Himanem, 2001)

Instead of having the *opportunity of changing* a part of our lives, as technology allows us to do, we feel often in fact *obliged* to change everything according to "technology" and not free to choose when running very fast, with the aid of machines, if we want or really have to do. We behave as if we *must* go on always running, as if someone were hunting us.

Difficult it can be called a "revolution" to be almost compelled every day to follow and learn new gadgets and devices, which we understand only for a little part, often use not at ease, waiting for new gadgets and devices in few years or even months which will compel us to learn them again, understand only for a little part, often use not at ease… Or it is just the usual realm of Market, stronger and stronger, the same that in the 50th and 60th of last century invited us to buy cathode ray tube television sets, washing machines and Bakelite cups?

Paraphrasing a text of the already quoted rock band King Crimson, it is possible that confusion "will be the epitaph" of the present often so called "digital" age, when "post truth" seems to be ruling if not everywhere, on many sectors of cultural, political and even economical life, and those who argue against "post-truth" often may have the impression of being "alone":

Post-truth has gone from being a peripheral term to being a mainstay in political commentary, now often being used by major publications without the need for clarification or definition in their headlines. ("Word of the Year 2016 is," 2017)

Though, in my opinion it is very difficult not to see the substantial non-existence of many of the topics which have stirred the debate in recent years, most of them gathering around the word "digital".

We live – despite of all the ads – in a very disconnected social context, where to divide learning from experience and even from thought is probably convenient for someone. It may be not by chance that, out of the described trends of the virtual world, in the real present the citizens of the earth are more and more giving up on a direct and responsible participation in the democratic life of their countries, delegating and entrusting their destiny to strongmen (or strongwomen).

All those "disconnected connections" put us in the center of a big contradiction, not so difficult to be seen and even forecasted yet at the beginning of social networks time: "Information technology can link geographically separated people and help them locate interesting or compatible resources. (…) they also have the potential to fragment interaction and divide groups by leading people to spend more time on special interests and by screening out less preferred contact. This paper introduces precise measures of "balkanization" then develops a model of individual knowledge profiles and community affiliation" (Van Alstyne & Brynjolfsson, 1997).

WHEN THE PLANS OF MARKET ARE WRONG

"They have designed all our future!" "They already know all the products they are going to sell to us during next twenty years!"

Words like these are often repeated and common people feel out of the possibility of doing something not only to eventually change the trend, but even to really choose what he or she would prefer from a production totally out of our control.

In fact, for example, we hardly know what 4K video is, and rumors say that within a few years 8K video is coming. We have just got accustomed to using a great number of passwords, and now the machines ask us for our fingerprints.

Nevertheless, also companies make mistakes, and we have many examples of how they have been not able, in the past, to draw the future.

The *LaserDiscs* and their dedicated readers were out from 1978. Unlike DVDs, they managed analog video, with better quality then cassettes, but with the impossibility of recording TV programs. During the Eighties, controlled from a PC, also some games appeared on such media, with all the scenes already recorded on the support and a cinema quality, and someone thought they could be the ideal means for interactive videos. Who has ever heard about them?

The *Digital Compact Cassette* (DCC) was proposed by a leader of consumer technology as Philips in 1992. Being fully compatible with the old format of musicassettes, it seemed to be the obvious medium for home recorded music now with a digital quality, for audiophiles already accustomed to audio

CDs, who on the same reader could have listened to both digital and analog cassettes. It was a total flop. The contemporary *DAT* system of Sony (with a non-compatible format of cassette, but superior quality) was in fact quite successful, but only among professionals.

Sony *mini-discs* were designed for a non-sequential digital home video recording. If they had some commercial impact, it was for a very little time.

In this case, the industry had apparently supposed an interest of consumers in the quality of music recorded in homes and the consequent need of dedicated devices, much more superior than the real. The matter concerned also the well-known problems of copyright when technological devices allow people to record and distribute protected music without permission and, anyhow, in order to see an actual diffusion of digital music not only on classic CDs we would have to wait for the iPod and MP3 readers to arrive on the market, at the beginning of the new millennium.

Mini DVD camcorders did not actually flopped, as they were sold since the second half of the Nineties in millions of pieces throughout the world, but we can say that they were, from the beginning, everything but the "future". Clumsy in hand, worse in recording than DV cassettes cameras, not good for editing, they seemed to be designed just as a way of luring semi-illiterate costumers with the "magic" of the laser writing: devices born to last just the time to be cool, and to be soon forgotten.

THE FUTURE IS DIFFICULT TO BE IMAGINED

We can call them anachronisms, watching only some "pictures", taken by heart from books and films:

Picture 1: In a lonely planet in the deep universe, far ahead into the future, before they can watch the images shot of mysterious presences, the explorers have to wait for the film to be *developed.*

Picture 2: In times when starships normally fly across the universe, computers have the same interfaces and graphics capabilities of those of the early '80 (when the film was made) and *CRT screens*, not even flat.

Picture 3: In the often here quoted *2001, A Space Odissey* (1968), the CPU of the super computer HAL is so large, that the human has to go and float inside it in zero gravity, turning off one by one the single elements which make up its electronic brain.

Picture 4: In another stone mile of science fiction cinema, *The Forbidden Planet* (1956), Robbie the Robot (the prototype of generations of toy robots!) had hands like rough grippers and a glass "head" full of big and visible *valves*.

Picture 5: In the cartoon *Futurama* by Matt Groening, one thousand years ahead, the robot Bender had to recover his vital instructions from a *floppy disk*.

Picture 6: In quite recent science fiction movies, like *Avatar* and others, people manage computers moving very quickly and skillfully hands and fingers on transparent touch screens, taking information like orchestra conductors, or expert guitar or piano players: very spectacular, as well as according to the most recent (of today, not of that future) trends of the market!

And so on.

Many times, on magazines, blogs, paper and electronic books, we read words by some people explaining where the "future" is going. Sometimes they are serious trials to understand which could be the possible evolutions of our society, ways of living and producing, also due to the technological progress, but many times they are very questionable opinions, not strong enough to stand a rational discussion, but often declared as they were the only indisputable truth. We have also to consider that, on the other hand, there are average interlocutors who, in the years, never have felt the need to ask, for example, what expressions as "2.0" and similar actually mean.

So, during latest decades many times we have read about the death not only of the pre-technological printed book (as we have already remembered, from the times of video cassettes!), but also of the mouse as pointing system on graphic interfaces, made obsolete from time to time by track balls, vocal commands, touch screens or other.

The new device, for definition, is better, is cooler, is newer (of course!), so the older device is to be replaced. Often regardless of actual eventual problems of use, as the writer of the article apparently does not care of them and probably, writing, he or she uses only the keyboard (as in all American films and TV series, moreover, where never people use a mouse working at PCs, but only keyboards, very quicky, for everything, without doing any mistake! Absolutely funny!)

Well, if we – normal people and not only former super skilled hackers hired by FBI – we assume to go on using sometimes computers and other devices not only to browse collections of photographs, but for example to give commands taken among hundreds offered by the complex software of word processing, graphics, video editing and so on, some very light and precise movements of a mouse on a horizontal surface are surely much more convenient than wide and clumsy actions on a screen, especially if vertical and large, not to speak about remembering just exactly (if not, the machine doesn't understand!) a huge list of typed or vocal prompts, hidden inside menus and submenus.

But to move arms and fingers around it looks so cool, as well as saying, in a Star Trek style: "Computer, do it!"

AN OLD FASHIONED CONFUSION RUNS ON THE WEB

At the times of our great-grandparents, to certify that something was true, the magic words were: "Newspaper says it". We have also heard, in more recent years: "Computer says it!", and it is evident that the familiarity with technology has not made common people less naïve, but possibly the ingenuity has even increased with which many tend to believe to everything is told not only on mass media – some of them should be influential, at least – but also on social networks, where news of every kind is often taken at once and shared as it is, coming by anyone, without any sort of control. Not only the so called fake news, created and used by someone in order to manipulate the reality according to various intentions, but even evidently playful pieces of news published on satirical web sites, whose only purpose is to make people amuse and laugh (and maybe consider important things from a lighter point of view) are sometimes taken reading only the titles and spread everywhere with passionate comments and endless discussions on line, as they were the center of the world.

This habit of shifting the attention from what really happens to what *is said* to be is not only a joke, but now affects many fields of public life, economy and politics. To indicate it, the new term *post truth* has been created which has been awarded as "word of the year 2016":

After much discussion, debate, and research, the Oxford Dictionaries Word of the Year 2016 is post-truth – an adjective defined as 'relating to or denoting circumstances in which objective facts are less influential in shaping public opinion than appeals to emotion and personal belief. The concept of post-truth has been in existence for the past decade, but Oxford Dictionaries has seen a spike in frequency this year in the context of the EU referendum in the

United Kingdom and the presidential election in the United States. It has also become associated with a particular noun, in the phrase post-truth politics. ("Word of the Year 2016 is," 2017)

Here *post-truth* is still classified as an adjective, but we could say that it is now a noun, the name of the other side of truth in our technological, virtual, "digital" and quite confused society.

There have been times when for an industrial, a commercial or financial firm it was enough "to be on the Internet" to have a big growing in the stock market. The so called "dot-com bubble" (or *dot-com boom, tech bubble, Internet bubble, dot-com collapse, information technology bubble*) happened in the last years of the past century, making the fortune and misfortune of many:

As a result of the rapidly-increasing usage of the Internet, many investors were eager to invest, at any valuation, in any company that had one of the Internet-related prefixes or a ".com" suffix in its name, leading to a stock market bubble. During the bubble, the valuations of companies in the quaternary sector of the economy increased rapidly. Venture capitalists, eager to profit on this investment demand, moved to raise and invest capital faster and with less caution than usual. A combination of rapidly increasing stock prices, market confidence that the companies would turn future profits. (Dot-com bubble, 2017)

The Internet bubble heavily collapsed between 1999 and 2001.

They do not say anymore: "Go to our web site!" as apparently it seems too difficult. They say directly: "Download our app!" for everything!

A curious trend of latest years, after the boom of smart mobile devices, is the migration from single programs for browsing the WWW to hundreds of apps, pointed just to that provider, that service, that shop, which fill above all our phones. This is the other side, but not exactly the same, of using for instance single task apps for graphic effects, instead of full working programs, that's not professional nor the best for literacy, but sometimes convenient for average people and also for the experts, who know how and why to use them and need to spare time.

But to have a plenty of apps not of my bank, my phone provider, my club, but of someone I attend only occasionally, is totally contrary to any logic. Devices get full of stuff taking room, power and memory, when we'd need only, instead, a list of web addresses, or simple names, to point to or to look up in search engines in case of need.

Besides, to offer services "app based" rather than web based is much closer to what happened in old "pre-technological" times, when every good or service had unequivocal sellers and buyers, than to the ambiguous and shaky World Wide Web age, when sellers and buyers can even exchange their roles and mix themselves (what a confusion with Ebay, or Paypal!).

But as people in general, and consumers particularly are afraid of change and responsibilities and have to be somehow reassured that they have not to engage too much in thinking and doing things, every new way to go back to the old dear commercial relationships is welcome.

At this point, if we take also in account the participation in social networks of billions of people, each as he/she wants and knows and without any need of a common technological literacy or skill, a true risk is that as the whole of society we lose completely, even before we have understood it, the sense of what Internet could be, of what computers are, and give up perhaps forever the hope of taking part actively in a more and more complex reality, leaving its development and organization entirely in the hands of few.

ALL WITH THE HEAD IN THE CLOUD!

I remember, before somehow I managed to shut it down, that every time I put a SD cart into my laptop, a message appeared like this: "I see that here are pictures. Do you want to upload them on the cloud? "

My answer was, obviously: "Heavens no, it's almost all rubbish!" But I guess that many say instead "yes", and I know also that saving data on the cloud in many cases is the *default* option given by the native software. In my humble opinion, this is quite foolish.

To be clear, I'm not against the cloud, I do use it, and this text itself is being saved on the cloud, as I write it normally on a desktop as well as on a laptop PC, occasionally I access to it also from a tablet and a phone, and to have interconnected devices which upload and download every new version is a very convenient way to keep always the work updated. The same with many other works in progress, or with documents I need to share with others, put on line in folders accessible from every member of a group. Also it can be a reasonable idea to put on line some important documents, if we are traveling and we are afraid to lose our device, or if we fear that thieves steal hard drives and back up keys from home.

But to let everything be uploaded without even checking what is worth to be saved or not, not only fill literally the virtual space of rubbish, but once more it takes away from us the control on our digital contents to give it to… we actually do not know to whom. Besides, the cloud space is not all for free, as over a certain threshold we have to pay, and we maybe would like to pay because we actually *need* a space on line, not because we put there every photograph or video we take, including those very bad, which should be better erased.

And what about back up on line, as default, even for people who do not know in general what a "backup" is? If it could be a temporary solution for mobile devices (they can be lost), or the individual choice of aware users who decide that it can be actually convenient for their own well-known purposes, generalizing such automatic procedures presents the serious risk of definitively entrusting the management of digital devices, networks and all the basic stuff of the information society to perfect illiterates, who have not even an idea of where their data are, the main and the backup ones.

By the way, warning, attention! It happened to me only once or twice, but it is enough to spoil days or weeks of work. Sometimes something can go wrong "over there" on the cloud, or also "here" on my local device, so that the cross-checking of data goes into confusion and inside a folder it is not recognized, for example, the text of this book, just when I am finally going, with a dreadful delay, to send it to my publisher. When folders are updated, everything being synchronized according to the latest check, the files not found cab be simply erased. So, if I have all my work also in many devices but all connected, there is the serious risk to lose everything.

There are several technical ways to avoid such a hitch, but it is not so easy for users now accustomed to find always automatic solutions. What we can do quite easily, is to save periodically renamed versions of our important data on external mass memories as hard or pen drives, and never to connect them to the network. Or, in the case of a sudden and unexpected reset of our personal organizer on Google and from there, with a knock-on effect, also on our desktop, laptop, mobile phone, tablet, *iPod* and game console all synchronized, a wise preventive measure is to keep also a paper planner.

From all the "pictures" above, and many others which we could recall to memory and reflection, it is clear enough - I think - how it can be often a nonsense to say "Internet" as if all knew what we are talking about, as well as taking, for example, broadband availability and web accesses as absolute indicators of the technological "progress" of a country. We can get from them an idea of the dimension of something, but how people actually stay on the Internet and what they do on line is probably more important, and too many times the quality of the use of the Web is really very poor.

Besides, attending environments as Facebook, Twitter, Instagram, iTunes, closed and "private" (with submissions, passwords, logins and so on), above all if approached from dedicated apps, it can have just nothing to do with the true experience of the World Wide Web. In those networks we stay after having signed a contract (though most after a while forget it) and must follow their rules. And apparently that is the preferred way of most people of enjoying technology, well framed into a "protected" space where they have to learn only few easy operations and doing the same things of ever: reading, writing, sending pictures, watching videos, and enhancing their communications with emoticons.

TEXT VS. DIGITAL AND OTHER ILL-POSED PROBLEMS

I have already quoted several times the definition of "digital natives" given to the youngest generations who grow accustomed to technological devices. Coming mainly from empirical observations and emotions and with a very poor scientific base, this definition has been taken by a substantial part of the public opinion as an indisputable fact, often raising a wall of incomprehensible incomprehension among generations, with a social damage that is difficult to measure. In fact, all what we know from the psychology of evolutionary age is in the common sense assumed as useless, because the habit of holding mobile phones would have made nowadays kids as Martians. Not naturally our sons and daughters, whose presence and problems are real and tangible in everyday life, but kids in general, to whom the society and the school often look at as through deforming lenses. On the other hand, "youth discomfort" is assumed as an indisputable fact too:

Until recently, there has been a largely uncritical acceptance of the digital natives discourse. The claims have been repeated by other researchers, writers and commentators, which has helped to give the discourse a sense of legitimacy. This phenomenon has been called the "snark syndrome". This is the idea (taken from the Lewis Carroll poem, The Hunting of the Snark) that if you repeat something frequently enough it eventually becomes accepted as fact. (Byrne, 1993, as cited in Bullen & Morgan, 2011)

Another recurrent topic – in articles, meetings, discussions on Facebook - is the supposed contrast between "textual" and "digital" culture, both assumed as other indisputable facts. That is the actual triumph of post truth!

We can begin answering to some simple questions.

Is it our previous culture textual? Why, how? The culture of literate people up to the middle of Twentieth century was textual indeed, but mass literacy in the world has happened possibly later, together with the spread of television, and television, more than "text" is probably the base of the "culture" from which everyone of us actually come.

Is it the present culture "digital"? What does it mean? That reading news or watching TV on a web page is a significant cultural leap? That words written on social networks are by people aware of technology and well skilled in using it? That to write or speak a message on Whatsapp means to own new alphabets? And do we have an overall idea of how they actually work the devices we have always in hand, or only we have learned some easy procedures and repeat them by heart, as also monkeys could do? Are we really able to deal with the huge amount of information coming from the whole of digital networks, checking and selecting them, or rather we usually take the first one we get as it was that true?

In general, how can we learn the right "competences" if even the instruction agencies have been able not yet to decide which of the new "alphabets" are basic and which not, hardware and software run more and more "automatically" and every sort of apps seem to free humans from the annoying trouble of really know something? I have repeated here some topics already touched in other parts of this book, as too many times we read and ear words as "digital citizenship", "digital school", which sound much as "post-truth" and in the end mean probably nothing, but affect in a significant way our attitude in social life. Many of us seem to think that all is already decided about our "digital fate", as technology was a one way road.

In the "The Digital Competence Framework" the European Commission's science and knowledge service "identifies the key components of digital competence in 5 areas which can be summarized as below:

1. **Information and data literacy:** To articulate information needs, to locate and retrieve digital data, information and content. To judge the relevance of the source and its content. To store, manage, and organise digital data, information and content;
2. **Communication and collaboration:** To interact, communicate and collaborate through digital technologies while being aware of cultural and generational diversity. To participate in society through public and private digital services and participatory citizenship. To manage one's digital identity and reputation;

3. **Digital content creation:** To create and edit digital content. To improve and integrate information and content into an existing body of knowledge while understanding how copyright and licenses are to be applied. To know how to give understandable instructions for a computer system;
4. **Safety:** To protect devices, content, personal data and privacy in digital environments. To protect physical and psychological health, and to be aware of digital technologies for social well-being and social inclusion. To be aware of the environmental impact of digital technologies and their use;
5. **Problem solving:** To identify needs and problems, and to resolve conceptual problems and problem situations in digital environments. To use digital tools to innovate processes and products. To keep up-to-date with the digital evolution" (European Commission, EU Science Hub, 2017).

Personally, I would add also "problem setting", but a great problem, in general, is that "digital" competences, in common sense if not in official statements, are too many times thought as somehow separated or even opposed to competences "in general", as our experience with paper and senses was part of another life from the experience with screens, and this is not healthy. Today all of us go directly into "digital" from our "normal" life (it is not as when one had to learn how to deal with punched cards!) and this means that we do not need a special "digital", but a normal culture and literacy to understand for example if and what we can do by ourselves, using digital tools, and what is too much for singles or a small stuff not very skilled and requires the intervention of professionals. That is just what happens if we approach technology with open mind and curiosity, as usually children and creative people do.

CONNECTED OR DISCONNECTED? THAT IS THE QUESTION!

Today, technically, many of us are always connected, via broadband and wi-fi or 4G having speed enough to watch and transmit videos live. We could say we are even "over connected", as the habit of taking everything from the Internet make us not to use a lot of the features which our devices can perform perfectly offline and exposes to the contagion of *malware* (short for "malicious software", called also viruses), because of a continual virtual promiscuity, perhaps unnecessary. Not to speak of the habit of always looking at our screens, not paying attention to the real people and world around.

But Narcissus does not watch the others nor listen to them. He is, for his nature "disconnected" (Recalcati, 2014).

Every time we use an image, a metaphor, to describe something, we have to remember that it is to have a stronger, a more suggestive idea of reality, but it is not *the* reality. That is true for "digital" school, which is not a school made of bits, and it is also true when we talk about our time as the "age of Narcissus". Narcissus was very handsome, and most of us are not, and neither we really love so much to mirror us all the time, *in the reality*. But the overall result of the way we use our means of communication, as powerful and portable as actually in the hands of everybody, to get so poor an amount of real communication among humans, that curious and quite worrying situation maybe it is easier to understand if we consider those means just as the mirrors where we like above all to admire the image of ourselves.

Narcissism is in culture, in politics, in social life, and the individuals, gazing mostly themselves, are sheltered from social judgment and able not to worry about the problems of the world. But looking only at their point of view, or of their group, they probably add problems over problems, so that the age of Narcissus, under the surface when not at first glance, is a very problematic time. If it is not an explanation, this can be a decent representation of some aspects of the present.

If Narcissus – as Recalcati says, from his experience as a psychoanalyst - is *disconnected*, that is just an image which can help us to understand how, in the present plenty of connections and means of communication, we find so hard an actual communications among individuals and groups. Means are connected, over-connected, but many humans are not, so that they are not really able to enjoy digital connections.

This also agrees with my point of view, according to which the main feature of nowadays technology, or better of the way we are consuming it, more than progress is *waste*.

But Narcissus is too busy in contemplating himself to worry about looking with a critical eye to technology. It is more convenient for him to pretend that there is a straight and safe way that, through the "digital" present, leads to the future. So, while mirroring every few months in renewed gadgets, he waits always for novelties, gladly resets his memory and willingly conforms himself to a one-way world.

WIKIPEDIA AND THE WAYS OF A SHARED KNOWLEDGE

Up to six in ten articles on Wikipedia contain factual errors. The number . . . shows just how unreliable it can be to use the online resource as a sole means of digging up information. (Oliver, 2012)

On the other side:

Wikipedia editors have voted to ban the Daily Mail as a source for the website in all but exceptional circumstances after deeming the news group "generally unreliable". . . . The editors described the arguments for a ban as "centred on the Daily Mail's reputation for poor fact checking, sensationalism and flat-out fabrication". The Wikimedia Foundation, which runs Wikipedia but does not control its editing processes, said in a statement that volunteer editors on English Wikipedia had discussed the reliability of the Mail since at least early 2015. (Jackson, 2017)

I remember once, looking up an entry in Wikipedia, that according to the Italian voice a certain family of insects were spread all over the world "but in Poland"! One reads and asks in wonder:

What the hell does it prevent those bugs from crossing the borders from Germany, Ukraine, Lithuania, Czech Republic and Slovakia to Poland? And what happened to those insects when the state of Poland, in other times, was localized much more to the east than today?

Sure there must be and error, so I go and see the English entry and I find that the Italian one has been translated from it, where it is written that those insects live everywhere on earth "but in the Poles"! That is impeccable, as the translator probably had wondered:

Which is the land of Pole?

Poland!

Even if since many years I deal also with insects, I do not consider myself an expert enough to write on encyclopedias, but in that case I had no doubt entering the voice and correcting it: an error less in *Wikipedia*!

That is the point. Talking of Wikipedia or other similar projects, the matter is not if there are errors. Of course, there are, as in every work made by humans. And we all can also consult on line a voice which is titled: "Errors in the *Encyclopædia Britannica* that have been corrected in Wikipedia". The matter is just that, if we find an error in such a shared, free on line project, *we can correct it*! This, compared to all the story of knowledge, is simply revolutionary.

Naturally, as the Daily Mail warns, we cannot take Wikipedia's voices as the absolute truth. But it is and it has been the same with any other book in any other period of the history, with news on papers, with reportages on TV. The warning on Wikipedia could be applied to every scientific publication, as we always should compare many and various sources, before accepting something for true. Some are more authoritative and trustworthy than others but, in particular encyclopedias, where the voices are so many, it is impossible that they contain no mistakes. And if nowadays easy copy and paste operations can be often a problem, indeed, as fake news and information are often widely spread without a control, we have to think that in other times, when faked printed information were written, we did not even have the chance to go to check and had to trust the book we had available, because in most cases we could not get anything more.

To have the possibility of writing together all the knowledge, from a pessimist point of view can be very dangerous, as any illiterate has the doors open to enter and do damage. From an optimistic point of view, any attentive and informed reader can see an error and correct it. And after many years of experience, it seems that generally the problem of fake news does not affect so much the popular Encyclopedia on line and its reliability is overall good enough:

In 2005, the peer-reviewed journal Nature asked scientists to compare Wikipedia's scientific articles to those in Encyclopaedia Britannica - the most scholarly of encyclopedias, according to its own Wiki page. The comparison resulted in a tie; both references contained four serious errors among the 42 articles analyzed by experts. . . . A study published in the Journal of Clinical Oncology found that Wikipedia had the same level of accuracy and depth in its articles about 10 types of cancer as the Physician Data Query, a professionally edited database maintained by the National Cancer Institute. The self-described "free encyclopedia that anyone can edit" has fared similarly well in most other studies comparing its accuracy to conventional encyclopedias, including studies by The Guardian, PC Pro,

Library Journal, the Canadian Library Association, and several peer-reviewed academic studies. Still, because anyone can edit Wikipedia entries, they "can easily be undermined through malice or ignorance," noted BBC technology commentator Bill Thompson. Vandalism of Wiki entries is common in the realm of politics. In 2006, for example, slanderous comments were added to U.S. Sen. Bill Frist's biography page; the IP addresses of the computers used to make the edits traced back to some of his political rivals' staffers. To counter such activity, Wikipedia places editing restrictions on articles that are prone to vandalism. (Wolchover, 2011)

We can say that the shared on line encyclopedia is just a significant proof of how our society could change from the traditional market oriented environment, thanks to present technological opportunities, if the attitude of people also changes. One meets something and, instead of only saying that is right or wrong (that is freedom of speech and expression, but does not change anything), he or she intervenes directly. That is the power of being at once, possibly, active actors and *making changes*.

I know that most people are quite lazy, but is it not that we risk to miss a great chance?

In this book, I often quote *Wikipedia*, well knowing that – as my reviewers advice me - "it's not a reliable source in academic context". But, with its lights and shadows, it has become for sure a central cultural source in our global society, looked up everyday by many millions people as a common reference as probably never mankind had before:

With 90 thousand voluntary active contributors and 365 millions readers, it is steadily among the ten most visited websites in the world and definitely the first "non profit". . . . There are more than 19 millions headwords in 282 different languages. (Spadavecchia, 2012)

Have Wikipedia's structure and inner workings promoted its astonishing growth and enduring public relevance? In Common Knowledge. . . ? Against a backdrop of misconceptions about its governance, authenticity, and accessibility, Jemielniak delivers the first ethnography of Wikipedia, revealing that it is not entirely at the mercy of the public: instead, it balances open access and power with a unique bureaucracy that takes a page from traditional organizational forms. (Jemielniak, 2014)

Maybe we can say that, as knowledge and information are nowadays more and more developed out of the traditional vertical flow from the academy to common people, through new horizontal channels spread worldwide, the point is probably just how to improve the quality of "common knowledge". It is a great, difficult, necessary job!

THE WORLD THROUGH THE "GOOGLE GLASSES" AND THE BUSINESS OF VIRTUAL SOCIALITY

Here we do not refer the Google Glass project of an "optical head-mounted display designed in the shape of a pair of eyeglasses . . . with the mission of producing a ubiquitous computer" (Google Glass, 2017). Here we are writing of the simple action that many of us repeat many times a day, of going on the Internet and search something.

There are actually (or theoretically?) thousands of *search engines* available, some of them are quite popular as *Yahoo* and *Bing*, or still survive since better times as *Lycos* and *Excite*, or are given together with other services on line. But almost all the inhabitants of the planet use and have been accustomed to *Google*, so that there is in fact a single company which, offering us a good service for free, controls, on the other side, the great part of our digital steps on the world.

Doing a research we can manage keywords, and we can also (who knows it?) deal with Boolean operators for more advanced results. But if for some reason the guys of Google decide to hide something, they can actually cancel it from the world.

Google, unlike Wikipedia (that is a non-profit enterprise) is commercial and very, very powerful. Giving most of its services for free, it has been rated in 2016 as the second biggest company in the world, thanks to the fact of having billions of people constantly browsing its pages, not only the search engine, but Maps, Translator, Gmail service, Google drive, Photos, News, Hangout. Google Plus, Blogger, and naturally YouTube. Not only on PCs, but also and above all on mobile devices, where Android, the operating system of Google, in the first quarter of 2017 run on the 86.1% of all the smartphones and the 61% of all the tablets sold to end users in the world.

The result is that "Google's share of the search ad market is supposed to grow to more than 80 percent by 2019, according to research firm eMarketer" (Townsend, 2017).

So, Google controls navigation and ads on computers and mobile devices, Google makes the rankings according to which names and brands are visible on line, Google even states now if my site is safe or dangerous.

I have nothing personal against Google, I do use their good services and tools. But the question is if we are wise to let a single company to get so much power, as the "market" should not be the only leading strength in the global society.

YouTube, a video sharing space born in February 2005 and bought by Google in November 2006 for 1.65 billion dollars, is told to be the second site of the world for visits, after Google itself:

Most of the content on YouTube has been uploaded by individuals, but media corporations including CBS, the BBC, Vevo, and Hulu offer some of their material via YouTube as part of the YouTube partnership program. Unregistered users can only watch videos on the site, while registered users are permitted

Figure 1. From 2009, Android has outclassed the competition in the market of mobile phones
Source: Adapted from Statista (2017)

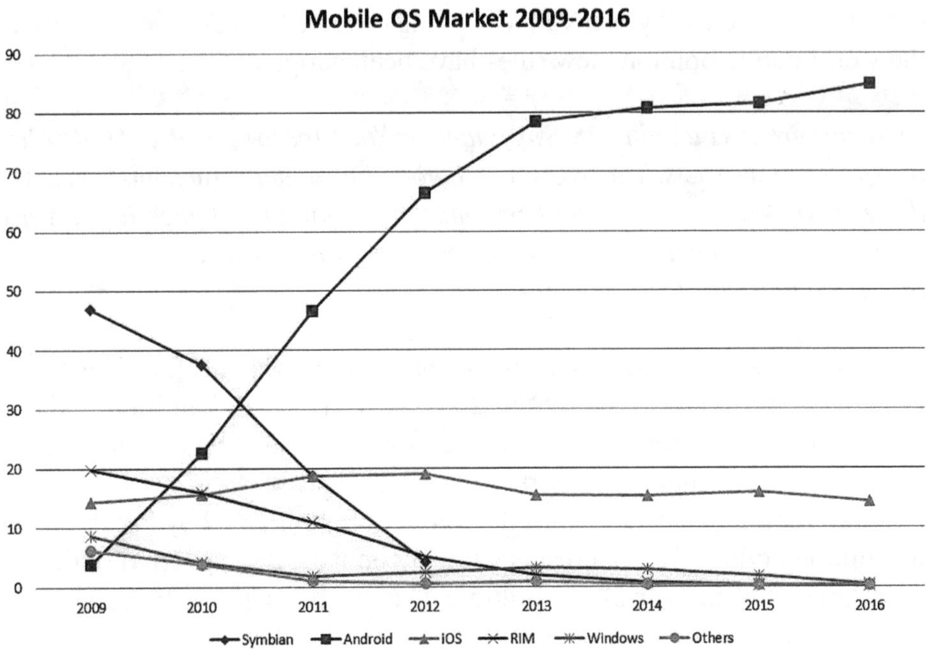

to upload an unlimited number of videos and add comments to videos. Videos deemed potentially offensive are available only to registered users affirming themselves to be at least 18 years old. (YouTube, 2017)

Registration is free, and people upload their personal videos, edited but also instant ones, even directly from cameras and smartphones (that's overall a mess, can we say it?), and also any kind of video taken from any source, parts of broadcasted programs, old TV and cinema. Copyright rules exist, mainly concerning commercial music, but in the end YouTube is probably the biggest file collection of the global information society, open to anybody, where we can find virtually anything. It is much attended by the young but, for instance, I personally met a group of old in a retirement home, using YouTube to collect and play the songs and the music of their young years, and listen, dance, enjoy with them.

Restrictions are limited to pornography (strictly banned from YouTube, though the web in general is full of any kind if pornographic sites, in fact accessible by anyone, regardless the age), but many have complained that the administrators of YouTube have not hindered enough the presence on its channels of videos by terror organizations, tutorials for preparing bombs, propaganda and recruitment messages and even sequences of murders and executions, watched by many before being removed. Under the pressure of the world public opinion, new rules have been adopted:

Extremist groups including ISIS use video on YouTube to recruit and radicalize prospective terrorists. This redirect method aims to deter such movements to drive "people away from violent extremist propaganda and steer them toward video content that confronts extremist messages and debunks its mythology," the companies say. (Snider, 2017)

On other social sharing spaces as *Vimeo* (independent, for videos) and *Flickr* (for still pictures, bought in 2005 by Yahoo for 35 million dollars), which offer a basic service for free and more advanced paid features, users can publish only their own works. *Instagram* is a very successful sharing platform for photos made with cell phones, bought in 2012 by Facebook for one billion dollars (!). Pinterest is a rather sophisticated system of indexing with "pins" and categories the multimedia contents on the web that we wish to share with friends.

I have quoted only some of the most popular examples of sharing platforms, in which anyone can take part, but hundreds exist and many new appear, commercial or cooperative, filling of ideas and opportunities the extremely moveable landscapes of the World Wide Web.

The first big occasion for common people of sharing contents on line was given with the so called "chat lines", at the beginning (second half of the Nineties) made by communities of computer and network enthusiasts, but soon ready to be attended by anybody, only choosing a nickname and a password and typing some text. There have been chats of every kind at once, oriented for subject, geography, age, cultural, political or sexual tendencies. With no any digital literacy required, a lot of people of every age began to meet other people in the Internet, virtually, but often not only, as that simple exchange of lines on a screen - perhaps unexpectedly, but it happens in reality - can move common interests, affinities, sympathy and even love, and bring the humans eventually to search one another in the real life:

Instant messaging offers two functions unique to computer-mediated communication (CMC): the ability to know who is connected to the shared space between or among friends, and the ability to conduct a text-based conversation in real time. . . . There is a positive relationship between the amount of IM use and verbal, affective, and social intimacy. Our findings are consistent with the liberation position of CMC relationships, and suggest that IM promotes rather than hinders intimacy. What is more, our research indicates that frequent conversation via IM actually encourages the desire to meet face-to-face. (Hu, Flower Wood, Smith, & Westbrook, 2004)

This a very large and complex topic, also related to possible dangers, of which the youngest in particular are advised to be aware, when the virtuality of the screens turns into actual meetings of real people, unknown.

Anyhow - in this case we can say for true words that often are used improperly - the chat lines and all what has followed have changed deeply the life of many people in present world.

Today, since several years, on *Twitter* and above all on *Facebook* hundred millions people create galaxies of different possible and actual uses. Intimacy, links, "balkanization" and many other eventual individual and social results of being always connected can be considered, studied, discusses. And recurring alarms are launched on the negative effects of social networks, especially on the young:

It's interesting to see Instagram and Snapchat ranking as the worst for mental health and wellbeing. Both platforms are very image-focused and it appears that they may be driving feelings of inadequacy and anxiety in young people," said Shirley Cramer, chief executive of the Royal Society for Public Health, which undertook the survey with the Young Health Movement. (Campbell, 2017)

But what we remember probably too little is, just as we have seen for Google, the often huge amounts of money flowing under all this sociality.

WhatsApp Messenger is a freeware and cross-platform instant messaging service, the most successful among other as *Telegram* (big files allowed, well encrypted, told to be the preferred for communications among terrorists) or *Viber* and others, following the way of an Internet based communication on mobile devices, more than via the old SMS, opened years before on PC's with *Windows Messenger* and *Skype* (good also for phone and video calls). But when in February 2014 WhatsApp is sold to Facebook, Mark Zuckerberg's company pays for it more than 19 billions dollars!

We can speak of technology, of sociology, of culture and generations; we can describe with enthusiasm the bright future that digital means make possible for all of us, and also clap to the success of some systems and

Figure 2. Whatsapp's outstanding success
Source: WhatsApp (n.d.)

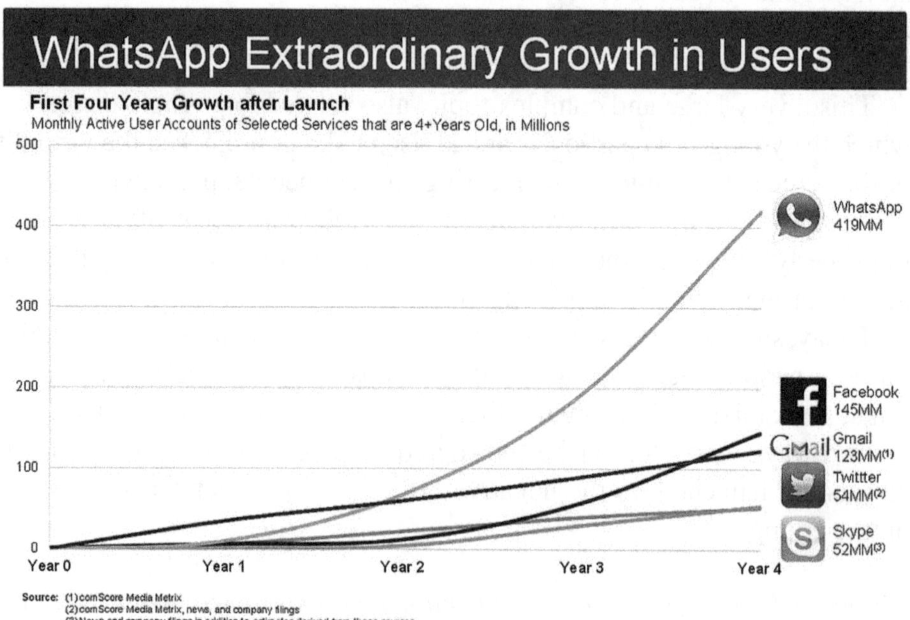

applications, so better than the others and so rewarded by the market, but we should never forget that the present season of social networks is moved above all by business. We have chosen in latest here, as a whole of society, to go along on the way of the digital era mostly according to the rules and the interests of few, huge, private and commercial global companies. We have maybe in part forgotten that the World Wide Web itself was a non-profit project, as well as still Mozilla (with its *Firefox* browser), *Libre Office*, the family of GNU applications and Linux operating system, and Wikipedia of course, successful non-commercial enterprises, which demonstrate that it is not necessarily the profit what always drives the progress.

It is not only a matter of entertainment, how to spend our leisure time, or to choose tools and platforms for our work and business. It is not only the question why, having at disposal for example a good and efficient free operating system as Linux, most users prefer to pay for Windows or Apple OS (addiction, lack of applications, overall difficulty, shyness of the free software to really make a different as well as strong perspective, need of a Big Brother)?

Today, even chairmen and prime minsters tweet and post of peace and war on social networks. That means that not only the channels of our private and public conversations, but also those of world politics are owned and ruled by commercial companies, quoted on the stock exchange....

Maybe, we should thing a little more of it!

IF TECHNOLOGY IS NOT A SEPARATE WORLD

Paper books and e-books, video games and hand puppets, insects in the garden and in augmented reality, paint software and crayons or finger paints, satellite navigators and paths remembered by heart walking mountain trails, real cars and driving simulators, real life and TV series of adventure, mystery, love... The list could continue indefinitely, as we live now in a world where everything is how as doubled and we are going to a future when “doubles” will be probably more and more “competitive” with the reality. But it does not mean that this is to be seen as a problem, or that we have to think – as many apparently do – that the virtual world is going somehow to overcome the real one, above all in the experience of the youngest.

If we replace fields and woods with factories or virtual replicas, spoiling the real environment, it is not a matter of technology. If we organize the time of our children so that they cannot run in meadows but only play at Sony, Microsoft or Nintendo consoles, it is not because of technology. The same if we set up school programs for computer coding rather than for nature watching, as no one healthy mind could ever say that digital devices are the future and nature is the "past"!

The anxiety to run behind technology is, in my opinion, the symptom of a bad relationship with hardware and software, as well as with some basic aspects of the physical and sensorial world. The idea that we have to adapt is a symptom of problems that, for our own reasons, not because of technology, we have with ourselves.

The contrast between digital devices and nature is probably only the latest evolution of the ancient idea according to which the mankind holds absolute power on all animals and plants and the nature in general and do not respect them, and is also due to the limits of our culture, which is separated, disconnected and inadequate.

So, sometimes we speak about the integration of digital environments and of the flows of migration of real people from the South to the North of the world as they were problems from two different planets. And we seem somehow to be oriented to imagine a future of integration between humans and machines, chips inserted in bodies, bionic limbs and cyborgs, more than to accept the integration of ethnic groups, cultures, religions of the peoples of the earth.

Problems always concerned in every process of integration or cohabitation are usually overcome if we see them simply as they are, part of a reality which is to be lived, with respect for ourselves and the others. That is generally what happens, when they are let free to know and express themselves under the supervision of attentive and interested educators, in the communities of children, who naturally and without many problems deal with technological devices and with more traditional and material tools, easily putting them all together, everyone at its right place, without forcing or pressure, as well as with other children and humans, regardless the color of the skin and the provenience. And every time something unexpected and pleasantly new happens, positively shifting from our forecasts, a little more ahead.

There is something deeply wrong, it is the symptom of a great insecurity, if we prefer to think the reality, present or "future", as a one way road.

SCATTERED PIECES OF A TANGENTIAL CREATIVITY

During the stage of the Tour of France 2016 to Mont Ventoux, there was an accident with a motorcycle and the yellow jersey Chris Froome was seen, broadcasted worldwide, foot running on the climb, looking and asking for a bicycle: a very unusual scene. After a short time in the Internet several pictures appeared, among which I remember one where the British cycling champion was on an athletic track, racing next to the Jamaican legend Usain Bolt! (La Gazzetta Dello Sport, 2016).

A famous video posted and "changed" by many during several years was that showing the Cuban "lider maximo" Fidel Castro falling down on a stage at Santa Clara in 2004. A well done version of it shows, coming from behind, the French soccer champion Zinedine Zidane who knocks down him with a header, as it actually happened, in the reality, when the Italy defender Marco Materazzi was head-butted during the 2006 World Cup final in Berlin. It is absolutely clear that it is a montage of two different videos, but the intent was just to make something funny and satirical, as individuals and members of web communities often do, when some facts from the world give them ideas and they can get the rough material (in these cases videos and pictures) to be put together and processed. That is an amusement, as well as an exhibition of technical skill ("Zidane Headbutts Fidel Castro", YouTube).

Digital photomontage with a good software is easy enough for still pictures and not so difficult now for moving ones, if we do not need to achieve the accuracy of professional productions. It is a way of sharing bright ideas in the web and the social networks, where also funny and ironical sentences, jokes, pieces of text from the most diverse sources, images of every sort, sometimes voices and sounds and a lot of music are sent all across the world by million people. When it is for fun and enlights emotions or reflections, all this works very well. Creativity, irony, good humor and information spread and sometimes *go viral*. It is a very new way to communicate one another but also to masses, to friends and strangers, potentially worldwide, impossible until a few years ago and in the hands of average people.

That could also mean a possible, incredible as unexpected *power*.

But when we "get serious", many things seem to change.

Though studies "found that people who consume more news media have a greater probability of being civically and politically engaged across a variety of measures" (Wihbey, 2015) and there have been indeed documented effective uses of social networks during political crisis by activists, militants, witnesses (but also by the Turkish President Erdoğan, who apparently crushed a coup in July 2016 by sending messages from his mobile phone!), reading the tweets of politicians as well as the discussions on line of common people about work and unemployment, culture and illiteracy, overabundance and hunger, peace and war, we do not get the same general impression of smart participation and innovation we have remarked for funny subjects.

Beyond personal opinions and points of view – these are kinds of data which hardly can be examined statistically – we can easily agree that most of the animated discussions on Facebook, where many people come in writing and arguing about the crisis between USA and North Korea, or off shoring in industrial productions, or computational thinking, not only do not change the core of those problems, whose solutions are matter for politics or economy, or for specialists, but also generally leave all the participants exactly of the same opinion as before. That is to say that, probably, there is not a real communication and those discussions are almost useless.

Moreover, after all the words spent by a lot of people, in every mass and personal media, on serious and complex problems such as unemployment or migration, which would probably required articulated solutions to which a good and open debate could bring contributions, many times (much more than in the past) what is proposed by politicians and approved by the citizens are in the end very simple and "rough" formulas, expressed in few words: no irony, creativity, good humor, as well as no perception of any power for average people, who more and more seem to perceive the impossibility of really decide anything.

It is not a problem *given by* social media, as its roots are more deeply hidden in many sectors of the society, but in social media it is clearer and evident that when we have the occasion we like to laugh of the world sharing with anybody, but at the same time we do not believe more in general that we can actually affect the political, economic, cultural choices.

Maybe it is not always exactly so, but this is a quite general perception, and the way according to which many move, act, and give up in the end on the hope of changing something in the world.

Creativity in this context is assumed as tangential, optional, "not necessary", and social networks are the ideal environment where we can stay "creatively" all the time and take a bright part in every discussion, without in fact doing anything real. So everything is lost into a sterilized box from which only a faint echo emerges of what we think, we do and we are.

REFERENCES

App Startup. (2017, April 18). *Mobile Operating System Market Share*. Retrieved from https://ids.uni.edu/theappstartup/2017/03/24/mobile-operating-system-market-share/

Bullen, M., & Morgan, T. (2011). Digital learners not digital natives. *La Cuestión Universitaria*, *7*, 60–68.

Campbell, D. (2017, May 19). Facebook and Twitter 'harm young people's mental health'. *The Guardian*.

Dot-com bubble. (2017). Retrieved on September 14, 2017 from https://en.wikipedia.org/wiki/Dot-com_bubble

European Commission, EU Science Hub. (2017). *The Digital Competence Framework*. Author.

Google Glass. (2017). Retrieved from https://en.wikipedia.org/wiki/Google_Glass

Himanem, P. (2001). The Hacker Ethic and the Spirit of the Information Age. New York: Random House.

Hu, Y., Flower Wood, J., Smith, V., & Westbrook, N. (2004). Friendships through IM: Examining the Relationship between Instant Messaging and Intimacy. *Journal of Computer-Mediated Communication*, *10*(1).

Jackson, J. (2017, February 8). Wikipedia bans Daily Mail as 'unreliable' source. *The Guardian*.

Jemielniak, D. (2014). *Common Knowledge? An Ethnography of Wikipedia*. Stanford, CA: Stanford University Press.

La Gazzetta Dello Sport. (2016, July 14). *Tour 2016*. Retrieved September 12, 2017, from http://www.gazzetta.it/Foto-Gallery/Ciclismo/14-07-2016/tour-2016-froome-vs-bolt-foto-piu-ironiche-web-160353429436.shtml

Oliver, A. (2012, April 19). Iffy-pedia: Up to six in ten articles on Wikipedia contain factual errors. *Daily Mail*.

Snider, M. (2017, July 20). YouTube redirects ISIS recruits to anti-terrorist videos. *US Today*.

Van Alstyne, M., & Brynjolfsson, E. (1997). *Electronic Communities: Global Village or Cyberbalkans?* Cambridge, MA: MIT Sloan School.

WhatsApp. (n.d.). *WhatsApp, The Anti-Marketing Growth Phenomenon.* Retrieved September 11, 2017, from https://growthhackers.com/growth-studies/whatsapp

Wihbey, J. (2015). *How does social media use influence political participation and civic engagement? A meta-analysis.* Journalist's Resource.

Wolchover N. (2011, January 24). How Accurate Is Wikipedia? *Live Science.*

Word of the Year 2016 is... (2017). Retrieved on September 14, 2017 from https://en.oxforddictionaries.com/word-of-the-year/word-of-the-year-2016

YouTube. (2017). Retrieved on September 14, 2017 from https://en.wikipedia.org/wiki/YouTube

ADDITIONAL READING

Recalcati, M. (2014). *L'ora di Lezione, Per una erotica dell'insegnamento.* Torino: Einaudi.

Virilio, P. (2010). *The Futurism of the Instant: Stop-Eject.* Cambridge, UK: Polity Press.

Chapter 5
Consumption and Production:
Culture and Food Between Global and Local Management

ABSTRACT

Today, though often we do not know, all of us are producers of information. The social web has changed the communication at every level, but participation is now so easy and so far from really advanced competences, that the technological awareness of people in latest years is possibly, overall, decreasing. Connection of media rarely match a correspondent connection of the human minds, let free from the trouble of choosing and less able to make decisions, and we probably should have to look less at the market and more at real people. Technology could give everything we need to be active actors on the stage of the world, but – this is true for culture as well as food - we are as at a crossroads where we can decide if to be active managers and producers of an important part of them, or buy everything, food and culture, from the multinational companies.

INTRODUCTION

Participation or Passivity?

Long, long ago, before social networks, first PCs were much less powerful than today but, when switching them on, people, if not engaged in a specific job, often get used to explore their possibilities, trying and learning software,

DOI: 10.4018/978-1-5225-3770-0.ch005

looking for solutions and possibilities, playing somehow with the machine and occasionally discovering something new and exciting. PCs were devices which we enjoyed to use without need of a network or connections, and this was the great difference from television, as well as from the computers of the previous generation (mainframes), normally accessible by common people, out of the headquarters of companies and public administration or the processing and research centers, only through a network and video terminals.

Nowadays, when switching on a PC maybe a million time more powerful, if not obliged to do a specific job, people go on Facebook and, if for some reason the connection does not work well, they feel disappointed as the machine could not be used. PCs are approached by many as a sort of very selective, individualized, interactive television.

The big problem of the web is television: that is to say our habit as society, since too many years, to spend a great part of our time seating and watching broadcasted programs.

Television is a great invention that helps us to be informed, to relax, to enjoy the stories, the music, the sport we like. But it has a dark side: passivity.

The web, on the contrary, would be a place in which everyone can take part actively, but whole generations accustomed to television go on it and, as usual, simply watch. Or also, many of them write words in Facebook and Twitter and even upload pictures on Instagram, but hardly one could say that this is to use in an active way the power of the web.

Smartphones and tablets probably are so successful because better than PCs they match the old habits of TV addicted users. Jaron Lanier, a pioneer of virtual reality, about latest generation of electronics devices writes openly of passivity, pointing out that it is also a matter of freedom:

Unlike a personal computer, a tablet runs only programs and applications approved by a central commercial authority. You control the data you enter into a PC, while data entered into a tablet is often managed by someone else.... When you carry around a smartphone with a GPS and camera and constantly pipe data to a computer owned by a corporation paid by advertisers to manipulate you, you are less free.... Citizens in the information age have to learn to be more than just consumers; they have to learn to be a match for their own inventions. (Lanier, 2013)

Many say also that the times are changing because the young, by now, more than watching television go on the web. Yes, but what do they mainly do, in the end, on the web? They listen to music, play games and watch television!

ALL PRODUCERS OF INFORMATION, NOT ONLY BLOGGERS AND "YOUTUBERS"

Every statement here is not to be taken as a piece of truth, as we are describing trends and examples and we could not say, as often many do, the young, the old, the web, the PCs, the mobile phones, the beef eaters, the Vegans, as if everybody agreed to a univocal meaning of these words. Even less in times when people usually listen very little to the others and all seem to go on with their own beliefs.

But we have to put our speech over something, so that we use words to approach and suggest a comprehension of reality, well knowing that real people and also real things hardly can be framed in abstract definitions.

Some definitions and words, though, are more closed to the real world than others.

So, when we say that all today are *producers of information*, this is very near to the truth.

Obviously, we are not all CNN or BBC, we do not have all the power and the influence of the men of the media as Rupert Murdoch or Silvio Berlusconi. But when we post a simple photograph on Facebook, Instagram or Flickr, we take our small part in the global system of information, all across the planet. It does matter if we do not consider our poor video on YouTube as a serious thing, if we think to address only a joke to a group of friends: putting on line words, opinions and every sort of documents, all of us are as true publishers. Good things we put visible on line raise the overall "cultural level" of the web, and bad things lower it, however small and humble they could be. But we have to consider the combined effect of millions, billions of them.

Then, some pictures and videos go much ahead of our forecasts and go "viral", even when we would not like it and try to draw back, to erase them. We cannot control the web, or better we can control our spaces, web sites, blogs, social pages, but we cannot do anything when something is taken from there by someone else to any other place, and virtual places are infinite. Many of us, the young and the less young, do not seem to be wholly aware of all this.

It is among the private and independent bloggers and "YouTubers" that we can find the most conscious and advanced, even if not professional, users of the information network of the Internet. New media are given for free by many providers and they are extremely easy to manage. So anyone can be motivated (LungHsu and Chuan-Chuan, 2008) to put his or her content and inform circles and communities around, but potentially all the world, not differently from what the professionals do. And some, even without means or money, become famous, demonstrating that another "journalism" is actually possible, in the time of the social web (Friedman, 2010), as well as the spread of new forms of "private reflective journals" (Foster, 2015).

But it is curious that in latest years, instead of thinking well of the huge democratic possibilities opened by this kind of information, instead of engaging in improving and developing it, mass trend seems to be generally to use less the blogs as a space of participation and move every discussion directly on social networks Maybe it is because people are not at ease having to deal with a too fragmented environment, or simply think that, compared to an "article", a tweet is more up to date. And if, for example, I quote a post of my blog on Facebook, people go and read on the blog, but then they write all their comments of Facebook. Mark Zuckerberg thanks and all ideas and opinions about that post, instead to be available in the time together with the original text, in few days are lost forever in the huge ocean of the social network.

Probably, elders are more blogging oriented and young more YouTube oriented, but sure all are Facebook oriented and there are no significant differences parting generations. People of every age come to express themselves on line along personal ways and from personal stories based on the common experience of audiovisual languages in which we all are involved. So, from my blog and web site of aged man who deals in schools with insects (Beneventi, 2013), it can happen to me to come across the YouTube channel of a teen-ager who take films of himself and snakes. Nice to meet you!

We only mention here and do not discuss the subject of how blogging and in general publishing on line, just enjoying the opportunity that everyone nowadays has to do it, takes average people (and institutions and lawyers, on new and unknown fields) to deal with problems once affecting only professional operators of the information, as copyright, fake news, possible legal responsibilities. Though actually only a decreasing minority of people is involved in this (another step towards passivity?), as the majority are unaware producers in the global, huge, diffused flow of the information in social and personal media.

MEDIA, MARKET AND AWARENESS, USING AND WASTING

To know all the history is not necessary for doing everything well, but to know a little it can be very useful, as we can compare the tools and products of now with those of yesterday.

Today, many devices for consumers and amateurs have features which make them suitable also for "serious" productions. Digital pictures and videos with cheap cameras and phones have the same resolution of those taken with professional means and, in good light conditions, not so worse a quality. Personal computers for everyone can be used for almost everything, with a good effectiveness, if not reliability. Easy software also available for free on line allows common people to design by themselves impressive web sites, with advanced graphic and tools.

But amid a market where, knowing what to look for, one can find almost everything, the majority of consumers are informed only about very few of the hardware and software possibilities and are addressed to buy and use devices and programs according mainly to brief and changeable popular trends, without being asked to give a real feedback if not in terms of number of sales. So that the development of many products not only for "niches", but also for possible masses of users as they could be for example those for schools, is in the end conditioned by the times too short in which they are actually available on the market before they are replaced and by the lack of information (not advertising) about them. People have not the chance to know, to test, to choose among them. And many times, the so called "innovation" is only in the direction of catching the curiosity of unaware consumers, with many functions which "taste" of future, more than to be really useful.

In the end, we know only very little of the "technology" around us and a great part of it is destined to pass without leaving any trace in our lives, an overdose of products cheaper and cheaper and more and more worthless and temporary, with a huge in-built human knowledge whose only goal is to be sold.

Waste is the "trade mark" of our time, when also technological rubbish is going to fill the planet, leaving a feeling of emptiness and dissatisfaction in its compulsory consuming inhabitants.

I like to remember when, for the one hundred years of cinema, we went into a kindergarten and let the children play with a real black and white 35 mm film in their hands, looking at it, touching, trying to see across all those very small pictures what was still and what moved…

In 1995 the experience of movies was strongly connected with physical supports as video cassettes. But it was already – we could say - a half materiality, for the vision of a movie, from a cassette as well as from another material object as a DVD, is not possible without machines which read those supports, taking out pictures that otherwise are not visible.

To take the pictures of a film in the hands and to see them directly with the eyes, before asking help to a machine to make them run under a beam of light, is something very different, more engaging and exciting. And to know what there was before, where we come from, helps us to understand a lot better what we are and we have today, possibly establishing a direct link with the senses that is the base, in the end, of the real understanding of what machines are and do for us.

In this perspective, much of the old stuff should not be considered as a trash, neither only a matter of nostalgia, but as something beautiful, interesting and useful to be known and managed, especially in educational contexts.

Anyhow, in that kindergarten in far 1995, when it was the time of revealing the small 35 mm projector we had taken, which so small, however, could not be, a teacher asked: «Where the cassette has to be inserted?"

THE UPSIDE DOWN WORLD OF "TECHNOLOGY" PUT PEOPLE UNEASY

Many years ago I was holding a course on computers with a group a teachers of primary school, in Italy. It is true that in my country we have always been dealing with technology in education in a rather conflictual way, but what it happened is typical.

I said, by the way, referring to the most popular operating system on PCs:

Actually, it is quite strange that in order to finish our work on Windows we have to push a button on which it's written 'Start'…

I saw their astonished faces, as to say:

So, you are meaning that using a PC I can go on thinking with my usual mind?

What I have to do?

I have already written about this wrong question, if talking of computers, which basically are multi (very multi) purpose tools, good for a huge variety of works. What we have to know well is just our work, our own purpose, and then *we* have to understand, eventually with the help of someone else, how a PC and its peripherals, a local network, the whole World Wide Web can be useful in doing it. In other words, the focus should be always on the work, not on the tool. If not, people find themselves uneasy.

But humans, when not driven by a specific job, "learn" well computers also if they can play with them, exploring ranges of possibilities according to their interests. This is why PCs are more a stuff for the "young", for the simple reason that the young are generally more used to play than the old people. But PCs are also stuff for the artists, the scientists, for the creative, more than for the narrow-minded people, who do not like to explore unknown worlds and to take responsibilities.

This is also why a "narrow" and bureaucratic use of technology put often us uneasy, somehow compelled from the exterior to think and behave in a different way of our normal and usual. And this is probably why other devices which actually are computers at all, as tablets, smart phones and video games consoles, are accepted by average people in their lives with more confidence and less fear than real PCs.

Something indeed is to be learned, very basic, when approaching to use computers and digital devices in general: how to switch on and off the machines (not too many lessons on that, please!), the use of icons on the screen and inside programs, how to save and load data... But all these things, we learn by doing, almost without realizing it, simply if we apply the use of devices on what we do.

This is the right, visible side of the moon, from the point of view of humans, able and free to use digital tools in their activities, which exist (the activities) also not depending from the tools.

This is what too many times, during too many years, people have not done, going on teaching and learning "the use" of computers, regardless of what they actually had to do with them. And because of the dominant position of a single company, leader in the field of office software, we have continued to think that we should learn "first" the office programs, even many years after PCs had finished, in the experience of most people, to be mainly office machines and had converted into multimedia ones available in every home, or in artistic studios, used for entertainment and gaming, video, photo, music applications above all.

Due to this ideological idea of PC, also some weird training programs have thrived across the world, such as the *ECDL* (European Computer Driving License), that even in present time teach people that, for a proper use of PCs, spreadsheets are "basic" and video, photo and music are "optional". Just as in the real life experience of anyone, born and growing in the digital age dealing every day with the functions of Excel!

This is what I call the upside down world of "technology" (in quotes).

Meanwhile, for other technology than PCs, for tablets and smart phones (which are computers at all), no much professional training is planned around (as few courses are on the use of television sets or microwaves). That is to say that the main media through which almost every woman, man and kid take part in the information society, such as television and phones, are reduced yet at the rank of household appliance.

PEOPLE, TV, VIDEO AND OTHER DEVICES: A DISCONNECTED APPROACH

My old Nokia N8, at its times, was surely on of the first mobile phones good for taking videos. Though it was rather small, compared with nowadays devices (3.5"), they had found the space to put in it a mini HDMI port, so that users could connect it directly to a television set.

Quite normal! Every device usable for home video, always, from the first Video Tape Recorders, to analog and digital camcorders, to photo cameras with video functions, all had been put on the market with their cables for an easy connection to a TV screen, or at least with a suitable port: AV, YC, HDMI. So, when mobile phone had become "officially" a device good for home video, it should have seemed a natural evolution to provide also for them that kind of connection.

Probably not so natural, if the mini HDMI ports has soon disappeared from phones (and it is rare to find them also on tablets) and its possible substitute, MHL connection (Mobile Hi-Definition Link), allowing to connect mobile phones to the HDMI of TV sets through their USB port and an adapter cable, after having been implemented in several devices for years (without any information for users), is now no longer present in most recent devices (as probably "market" has *not liked* it).

We are here as at the theater of the absurd.

Devices are sold good for making videos, but an interface to connect them directly to bigger screen is not provided. Maybe it is because - I try to hazard a guess - a mini HDMI port does not allow the new phones to be in time thinner and thinner, as the users like (and being also always larger and larger, if people put them in a rear blue-jeans pocket it happens that they bend, but probably also this the users like!). Almost nobody knows that in the USB port, (Universal Serial Bus, so universal that, having to do everything is often overused and crash, so that it begins difficult even to charge the phone) a function is possibly implemented for connection with TVs and users simple get accustomed to think that a connection between phone and TV is not possible. In the meanwhile, they take the nice habit to shoot videos vertically, which would appear immediately wrong in a TV screen, so that probably it is better to avoid any comparison and leave them in their blissful ignorance.

At this point the producers come to the conclusion that a connection between phones and television is *not interesting* for users and cut it completely off, as it is *useless*!

FREE US FROM THE NEED OF CHOOSING! TOWARDS A "LOOK BUT NOT TOUCH" WORLD?

Something similar has happened also with *LightScribe* discs, CDs and DVDs. Who knows about them?

LightScribe systems allow users to burn also the other side of a digital disc, the one without data, with a picture. As the print is made by a laser ray, colors are not available, but the surface of the disc appears as engraved and it is an elegant alternative, for works that we like to present well, to the usual inscriptions with a marker pen, as well as to the kits for applying manually labels to discs or to Direct-on-Disc Labels, which need a printer suitable for welcoming the discs with pre-applied labels.

For a long time, *LightScribe* read-write drivers have been provided with almost every desktop and laptop PCs, but the majority of the users probably never have even imagined of their existence. I do not know it is a problem of my "peripheral" country, but I remember that I have been looking for years for LightScribe discs in the shelves of supermarkets, next to normal writeable, re-writeable and Direct-on-Disc Labeled ones: in vane! Maybe, meeting them by chance, someone had noticed them and eventually chosen to use.

After years, producers have come to the conclusion that people *are not interested* in *LightScribe* discs (see Figure 1).

To burn CDs and DVDs and to give them in a school, company, among relatives and friends, is a tangible sign of how anyone, thanks to nowadays technology, can convert into a *real producer of information* and, if we add a self made printed cover for the jewel case, our production on a digital disc can look just like the music or the movies of the greatest artists and houses we can buy in stores. Maybe many average people still would like to put their own productions into a bookcase in their sitting room, close and "equal" to the professional ones on the side.

As optical discs recently have become "old-fashioned" and people prefer (really?) to have nothing in their hands which could remember that digital age is possibly something more than typing texts on Facebook, I give here to examples, of scenes with kids:

1. Students and their teachers in a school make their own video and publish it in a DVD. After taking shots, they edit and divide it in scenes, then choose also something from the backstage, cut amusing sequences for the "extra contents" and organize a menu from which the audience can go and play the main film or the others. They design the label for the disc, a round picture with a hole in the middle and essential text, and a more complex graphic project for the jewel case, paying attention to the layout and the right dimensions. At the end, a product is given to parents and friends which hardly they can distinguish from the "true" movies they buy in stores for their Blu-Ray player, by Disney or Pixar, Warner Bross or Paramount. This is from the "past";

Figure 1. Lightscribe: This is what we can do on Lightscribe CDs and DVDs
Source: Lightscribe (n.d.)

2. Kids (but also women, men, young or old people) in their leisure time take video shots with their phones and, without any editing, upload them directly on their personal spaces on social networks. While at school they study on tablets often digitized versions of school books from the Eighties or the Nineties and probably do not even have a YouTube channel of the institute, because to show kids on line is dangerous. This is the "updated" present!

In the two decades across the old and the new millennium, in fact, many possibilities had been shown on the shelves of supermarkets and had been also experienced in schools, but maybe too much on the side of *production.* This probably caused confusion in the minds of users accustomed to think themselves mainly and only as consumers, as well as in educational institutes, worried about preserving the traditional student teacher relationship.

As to draw and print by themselves labels of every sort, business cards, posters and T-shirts, as well as to burn at home or school communication objects as CDs and DVDs, all this probably means, for people who do it, the trouble of knowing and thinking, somehow playing and amusing, trying and maybe discovering something new, enjoy in being veritable makers. Too much complicate, for a market and a society more and more oriented to great numbers, standardization, appearance, always increasing the power of machines and decreasing the possibilities of choice, whose customers have not to know what they can actually do, but only what they can buy just in that moment, always addressed along new, not well known but above all puzzling paths. We have apparently to live across a "look but not touch" world, not to discover the power we have in our hands, if not really to change it, to take an active part at least.

3D printers are one the latest border products of that "old fashioned" idea of taking technology and being free of choosing what to do of it. They are too much new and innovative not to be shown and explored in fairs and meetings of production and education, but they are taken apart for the great public, left intended for specialized users (though sometimes the "specialists" can be the pupils of primary schools!). Using 3D printers, one has to think, to make projects, to consider shapes and materials and, above all, one expects to obtain some tangible products in the end, as well artisanal as hi-technological. They put users directly on the production line, breaking the border: a great responsibility.

Robots are probably in the middle, with their well rooted appeal to do things instead of humans (remember *Robbie in The Forbidden Planet*, 1956!) until we know that, on the contrary, most of them need instructions from the humans to act properly. It depends much on what kind of intention people put in managing robots.

Apart from industrial robots, which are designed to carry out very specific tasks in factories and are programmed by specialists, if longtime robots in houses like washing machines and less longtime as kitchen appliances do not affect much our imagination, the so called *robovacs* (robot vacuum cleaners) surely are much more future oriented, just as in the *Jetsons* world.

But Lego robots, for instance, before they can go and run and activate their tactile, light, movement, impacts, sweetness or acidity or whatever else sensors or microphones or cameras, claim always somehow the action of aware humans who supervise or even program them, who tell what to do "in case of" (Nguyen, 2011). Playing with them, a human being easily feels as the actors of another future.

Toy robots with animal or human shapes are also often interactive and programmable objects.

I have seen Bee-bots in the first grade of an elementary school acting together with children in a video play of *Pinocchio*, disguised as animals and with all their movements well drawn in the space, so that they could move and interact, also speaking with the voices of the kids, who "doubled" them.

I am taking part in some projects (waiting for being funded by the European Union) for another theater where human shaped robots dance with girls and boys, with the already quoted professor Lariccia still going strong. He goes ahead along the road traced from the Eighties, as in fact the first LOGO language by Seymour Papert, before going on PCs, was implemented in a programmable turtle robot, whose the Bee-bots, for instance, are a modern, cheaper and more popular version.

Lariccia says:

LOGO is not only a programming language, but also a very powerful metaphor of what the 'automata' are and what they can do for us, a small but important glimpse toward artificial intelligence. In fact LOGO language relies deeply on subjective geometry which is at the core of the theory of relativity. Children are encouraged to think themselves as automata. The so called "turtle geometry" is an essential part of the thought process implied by LOGO programming. (Lariccia Personal Communication, 2017)

For people, not only to look, but also to touch things is very important – even if they begin of "thinking themselves as automata" - for having a balance in their life, for keeping active the senses and the awareness of the natural part of the world. As there are downsides in always using technology, from which we have to defend.

Satellite navigators, for example, which are a great invention and help people to drive their cars and steps exactly just where they want, saving time, money and physical and mental energies, risk, along with trains, buses, planes in large spaces and elevators and moving walkways in the small ones, to remove from humans' experience the physical perception of the environment in which they move. So some people attend courses of orienteering, whilst others, especially the younger, are not able to go just out of their town without a GPS connection.

Visible signs in the territory and landscapes are important, details we can memorize and then recognize at a glance, close or from far, engaging our memory and emotion, everyone of us in his personal way.

For the kids and teenagers, whose personality, vision of life and relationships are still taking shape, to know that we do not need always technological prostheses it is even more important.

So, being allowed to choose which hardware and software we like or need to use, having the possibility and the time to know what is available around, to test what seems to be interesting, to go and easily take it, without in the meanwhile losing sight of the fact that we are human beings and sensitive, is a basic condition for a serene and active life of the individuals and the communities. As the focus must be on the humans, put at the center of our attentions and intentions.

It can also be interesting, by the way, to see towards which aspects of "technology" we are addressed by commercials in TV, where we are overall only marginally taken to think of robots in factories and digital controls applied in industry to test the quality of everything, from microcircuits to coffee beans, but showing, for instance, new cars – that are also by now largely digitally controlled, engine, consumption, trim, security, sensors and cameras – they remember to us and stress more the direct connection from the dashboard of that car model to social networks!

But if we shift our usual perspective and take the time of observing, and thinking, out of the trite models of consumption (which are the same in the end of pre digital age, but at ultra speed!) maybe we can glimpse, for instance, subtler but significant:

Real examples of a feed-forward ontology in which information begins to flow from the future to the present. . . . An American researcher has put in her ultra-old parents' home sensors that are able to detect the movement and the speed with which the two move around the house. Why? The answer is that, in some previous pilots, it has been verified that a slowdown in walking speeds of 5 cm/second implies an 86% chance of falling into the house in the next 2/3 weeks. (Mazzuchelli, 2017, loosely translated by Paolo Beneventi)

THE PESSIMISM OF IGNORANCE

There is a famous sentence by the Italian politician and theorist Antonio Gramsci: "I'm a pessimist because of intelligence, but an optimist because of will" (From "Wikiquote").

If we could synthesize in few words a glaring contradiction of our society, we could say: "We are optimist because of possibilities and pessimist because of ignorance."

On one side, technology enthusiasts describe our world as a nonstop show of new amazing devices and apps and services, which make our lives everyday better and smarter, as we can let them working for us, toiling for us, thinking for us.

On the other side, catastrophic and apocalyptic people see a world ruled by ruthless and heartless machines, where humans are losing every freedom and possibility of choice, forced to live under a dictatorship with no hope.

In March 2017, I was invited to be part in the jury of a film festival for the kids and the young and, speaking with the other members, we notices that the distinctive common trait of most of the movies was pessimism. Even in the most serene and light, death was always present, and also there were scenes which because of their violence never, only some years ago, would have been shown to a public of young.

Pedagogy also changes. I remember a page of the Italian version of *Childcraft*, the famous American encyclopedia for children, very widespread in my country during the Seventies, where a carnivorous dinosaur savaged an herbivore. Something dropped from its neck, but it was not just blood, as it was in fact transparent: "It is not to impress the child!" the presenter told me.

Following these criteria, today we should switch off forever all the televisions in the world!

But one thing is not to hide to the children that there is evil on the earth (they still know it, they feel it, and responsible adults have to educate them to face the reality), another, in my opinion, is to send messages in the direction of cancelling completely from their horizon smile and hope. That is, by the way, quite impossible, as always the young try, in their green years, to find something light, pleasant, amusing in everything, even in tragic situations. So, if they find also adults able to smile and laugh with them, it is important.

But why do I link pessimism and ignorance?

Ignorance is what we do not know, and it is the other, necessary side of the "optimism" of the market. If we knew things, or also if we tried to know things and considered knowledge as a value, we could not completely believe in this kind of market, where technology is proposed in the same way of detergents for laundry, which for forty years tell us always that whiter it cannot be. The same is for that microprocessor, that display, that graphic or sound card.

But looking at a glittering surface we do not know, that is to say we fear, what it could be behind. And in the same way we do not know, that is to say we fear, the human beings (other language, other skin or somatic traits, other religion) we are not used really to communicate with, despite of all the cheap, powerful and easy means of communication which superficially invade our lives.

Some visions of the future today by contemporary writers and authors of books and cinema are surprisingly similar to the visions of the future many writers and authors had in the Fifties of the past century. They were just coming from the terrible World War II and we are today in a situation which, by many, is considered not so far. These are Pope Francis's words:

Even today, after the second failure of another world war, perhaps one can speak of a third war, one fought piecemeal, with crimes, massacres, destruction. (BBC News, 2014)

The word that is being repeated often is insecurity, but the real word is war. . . . The world is in a state of war in bits and pieces. . . . Not a war of religion. There is a war of interests. There is a war for money. There is a war for natural resources. There is a war for domination of peoples. This is the war. (Millar, 2016)

Ignorance is the mother of all fears, of the humans as well as of the machines. And this is rather paradoxical, in a time when we actually have machines that virtually allow us to multiply endlessly communication, and therefore knowledge with anyone in the world.

A CHANCE TO RESTART FROM THE HUMANS AND THE HACKER ETHIC

On the Internet with the body?

Let me paraphrase here that old sentence from the Seventies, "At school with the body", as the starting point of a central topic.

But I could also refer again to *The Hacker Ethic and the Spirit of the Information Age* (Himanem, 2001):

A new work ethic that challenges the attitude toward work that has held us in its thrall for so long, the Protestant work ethic, as explicated in Max Weber's classic The Protestant Ethic and the Spirit of Capitalism (Weber, 1904). . . . The interesting promise of the new information technology is that it could make a new form of task-oriented work possible. . . . When the individual at work is still treated as a dependent receiver, a trend is encouraged in which leisure is also reduced to passive amusement, with no room for active passions. . . . Only when an active work model has been achieved will active leisure also be realized: only when individuals become self-directed in their work will they be able to become active creators in their leisure time.

We need our time back. We need the possibility of thinking and listening with all our beings and bodies, free and happy of dreaming, imagining. And in these times of virtual worlds and "second lives", with all the opportunities and new perspectives concerned with a massive use of technology in every part of our lives, it is more and more important for humans to keep well, in many cases to recover their natural roots and nature, their tangible reality, made of matter, senses, natural environment, animals and plants, flavors and smells. Not to run the risk to be brought too much outside into the dream, even more because in most cases it is not our dream, but it comes from television, video games, electronic charms and in the end from the Market. And when huge crowds of people are lead to dream the same dreams, all throughout the world, it is in their deep previous culture, if it lasts, in their spoken and written language and tradition, as a part of a nation or a community, as well as in their "material" capabilities as individuals and groups, those related to the air, ground, water, fire (we can quote here the "elements" of the past as a representation of the "eternal" feeling of human being, as love hate, fear, goodness, jealousy, through the eras and societies), it is in those basic human components that they can find the antidote against the overdose of "digital", against the general homologation to the flat trend of consumerism, all the more dangerous as probably in fact no longer feasible by now on a planetary scale and thus often reduced to empty ideology.

Confidence in their own body, active attention to the others and nature are to be saved and improved much more today than before, in a supranational and interconnected world where many old rules are over, new ones are not yet settled and rooted and many people have not economical, social, cultural, political bases on which properly organize their lives.

Sometimes, just from suffering and discomfort something new is born, difficult to be expressed with the words of the political tradition and often intended to be precarious and with uncertain results, as the movements Occupy Wall Street or the Arab Spring. More than to frame them into the old categories, and obviously refusing to trivialize them into a perspective of a "digital" incoming revolution, motivations and strength of these movements of humans are maybe to be taken above all from our living, as corporeal experience, as for instance Judith Butler does:

Assemblies of physical bodies have an expressive dimension that cannot be reduced to speech, for the very fact of people gathering "says" something without always relying on speech.... Butler links assembly with precarity by pointing out that a body suffering under conditions of precarity still persists and resists, and that mobilization brings out this dual dimension of corporeal life. . . . By enacting a form of radical solidarity in opposition to political and economic forces, a new sense of "the people" emerges, interdependent, grievable, precarious, and persistent. (Abstract of Butler, 2015)

Without necessarily being masters in critical social theory, or taking part in mass epochal movements, we can also begin from very easy things, from more care, for example, to that "active leisure time" of which Himanen writes. Not to passively accept always what is given to us, but simply to remember to listen, maybe a little more every day, to our senses; not to go almost automatically after the clichés according to the main trend of the moment, but to ask ourselves sometimes the question of whether the things so done work really well.

I have already quoted the situations when people do not believe to their eyes and do not see, at the cinema or in television, that pictures are warped because of a wrong format.

But also with sound and music, also at theater, many signals indicate something which could be called a general decadence of taste and judgment, to which we might react.

Songs in churches can be wonderful. In ancient ones, old themes from the renaissance and baroque time, eventually with the accompaniment of organ, are just the right thing in the right place.

But why, then, many go there and play and sing other kinds of music, with a bad amplification set, so that everything echoes and the audience does not understand any word? How can the musicians and singer perform, giving so bad a quality, even if they are good in singing and playing, and how can people stay and clap in the end, after having their ears so tormented for so total a cacophony? Apparently, in these social and rather sad happenings "the event is the message" and to say that something is wrong is not an option.

In many pop concerts, the audience sings instead of the singer. That - I think - is something different from emotion, from crying and express enthusiasm, from all that could naturally happens in a meeting between a pop star and his or her fans. I try to explain better. If someone, at a certain point in the middle of a concert, feels to sing with the singer, it is a possible reaction in an engaging situation. But if all, usually and normally, at the start of a song begin to sing all together, it seems to be rather a habit, a conformist behavior, which could have nothing to do with that artist and that concert, as it always happens with every artist at every concert. People pay a lot of money for attending a concert, and then neither they listen to their idol well. So, people pay and gather in a great number maybe not for the music, but *just to be there*.

In theater, some actors use microphones even in very small halls, losing without any reason the spatial dimension of the sound. They are many on the stage, or turning their back: who is speaking? They use microphones put directly on their faces, losing much of their possible facial expression on that side. Why? They are actors, and face is one of their main means of communications, not a microphone holder!

But some use to do in this way and so many use, as it's easier. The actors "save" their voice, the public accepts that "convention" and in every place it is the same, no difference between the perfect acoustic of an ancient lyric theater and the bad environment of a circus tent, all treated in the same standard way.

I remember at the beginning of a piano concert, in a baroque theater, an organizer speaking first into a microphone and explaining something about that evening, and then, during the concert, the pianist stopping playing sometimes and addressing some words to the public, just as he was, from his seat, without the aid of any amplification. His words were not shouted at all, but they were perfectly audible, better than the ones of the organizer.

Maybe, happiness (or at least a less boring and flat life) is also to be able and used to realize the difference between a voice speaking in a baroque theater or in a trade unionist meeting!

Need of staying together and to do something, anything in the end, as a group. Need to show ourselves according to recognized clichés, without care of the actual final result. Systematic waiver to quality, as it is something that no one ever really asks. We resign ourselves to these poor community rituals, when on the other side we provide ourselves, in our daily life, with devices and machines possibly giving us the best of video, the best of sound (what great pictures the new curved TV set! What a good bass response that headset!); when we have at disposal parks and halls for good music and performances with amplifiers and theaters and churches without need of microphones. As the world is not a set of serial, interchangeable pre-packaged products, good for every occasion and context.

FOOD AND CULTURE: THE BIG CHALLENGE

Today, at least in economies not too much imbalanced, people can choose what kind of food to eat. We can take natural, local and biological one, or trust completely in the hands of multinationals.

If it is undoubted that seven billion people across the world can't be all fed from the garden next door, hardly can we assert that monoculture, standard industrial dishes the same all over the world, transgenic food and patenting of seeds is the only way to the future. Food is not only sustenance, but culture, identity, tradition, and when a community is allowed to take care directly of what women, men and kids eat and drink, overall quality of life improves.

As I come from Italy, I can easily witness about typical dishes, recipes, wines, cheeses, particular, curious in every town and village, different one from another in the range of few kilometers. It is an incredible richness of the present time, not only an heritage from the past, if it true that many young – not only for the economical crisis, but for a conscious choice - are going back to work in the country and we also see in many parts of the nation, just at the dawn of the 21st century, a spreading production for example of artisanal beers – good quality and variety - standing well on the market.

This is the "first" world, indeed, a still rich country, in a Western Europe where people since many decades do not know what war is (and we spend a lot of time complaining!). But it is not written in heaven that the majority of the humanity has to struggle to survive.

The possibility of redistributing a part of the production – of food, but not only - and above all of the richness in the world is a matter of politics, which is the way how the humans live together in a society. In the end, it depends on our choices and, if we want, with nowadays technology and knowledge, probably we can decide much more than we usually think towards which direction to drive our present and future.

Culture is the same of food, or even more, because, as it is not matter of life or death, we can approach the subject with more serenity. And, as we have not necessarily to deal with material goods, which need to be picked up, stored, conserved, shipped, delivered, all could be extremely simpler.

Today, thanks to the spread of technology, we actually can choose if to do a part of the information by ourselves, or to buy everything from the multinationals. This is for the first time in history, since very few years, and most of people in the world are not aware of that yet. But, if we just know how to do it, we can!

On 22 August 2017, European Commission Antitrust stops the Bayer Monsanto merger and open an in-depth review on the affair. According to Coldiretti (the main Italian association of small farmers), the 75% of the market of agrochemicals and the 63% of that of seeds in the world are controlled by only three companies (see Figure 2).

Figure 2. Seed and agrochemical market
Source: Natural Blaze (2016)

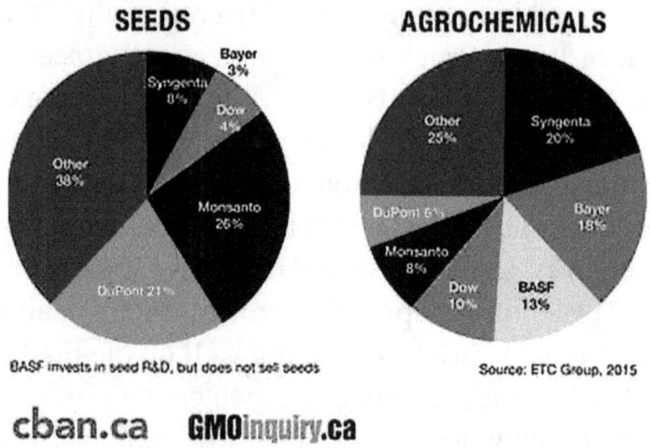

As widely reported, 10 big companies control the great majority of all the processed food production: "All the countries in the world are totally depending on a lot of processed food. Right from the milk to the meat. Every food item seems to be processed today" (Top 10 Best Food Processing Companies in The World, n.d.) (see Figure 3).

On the information side, Google (in the stock market now named "Alphabet", which owns YouTube) and Facebook (which owns Whatsapp) are in 2016 the second and the seventh industrial power in the world, competing with more traditional software and hardware brands as Apple (first), Microsoft (third), with innovative distribution brands as Amazon (sixth) and other companies dealing with energy, pharmaceutical, textile, telecommunication.

Comparing the list with that of ten years before, we can see at a glance how technological companies have grown in importance, but the most curious fact, probably, considering just Google and Facebook, is that they – as we have already stressed in these pages - have built their empires giving their service *for free*!

Figure 3. The biggest companies in the world
Source: Gray (2016)

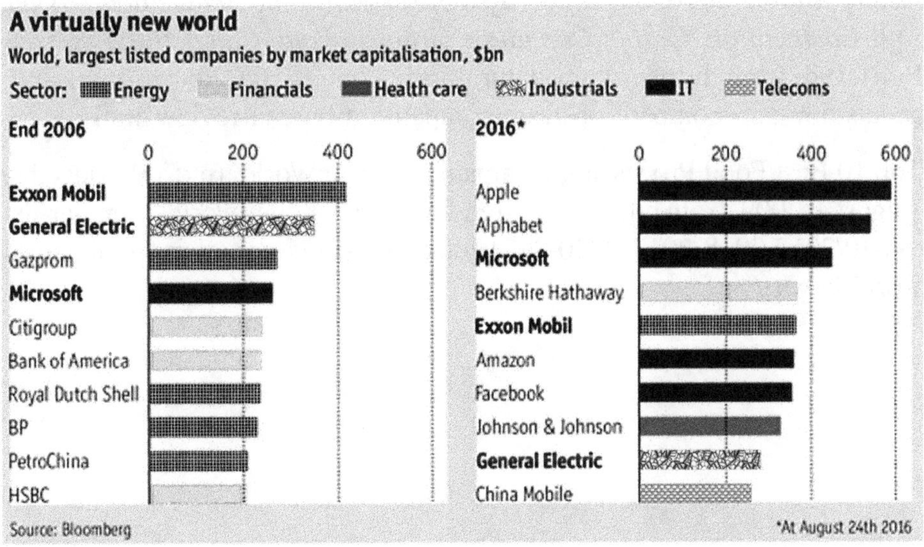

REFERENCES

BBC News. (2014). *Pope Francis warns on 'piecemeal World War III'.* Retrieved from http://www.bbc.com/news/world-europe-29190890

Butler, J. (2015). *Notes Toward a Performative Theory of Assembly.* Cambridge, MA: Harvard University Press. doi:10.4159/9780674495548

Gray, A. (2016, January 16). *These are the world's 10 biggest corporate giants.* Retrieved from https://www.weforum.org/agenda/2017/01/worlds-biggest-corporate-giants

Himanem, P. (2001). The Hacker Ethic and the Spirit of the Information Age. New York: Random House.

Lanier, J. (2013, November 27). Digital Passivity, We have backed ourselves into a virtual corner. *The New York Times.*

Lightscribe. (n.d.). Retrieved from https://i.ytimg.com/vi/0nh5ZvP59jM/maxresdefault.jpg

Mazzuchelli, C. (2017, March 27). *La tecnologia è il modo dell'umano di stare al mondo: difficile distinguere dove finisce l'una e inizia l'altro.* An interview to Cosimo Accoto. Blog "Solo Tablet"".

Millar, L. (2016, July 28). Killing of innocents proof 'the world is at war', says Pope Francis. *ABC News.*

Natural Blaze. (2016, September 15). *Bayer Monsanto Acquisition Triggers Anti-Competition Fears: Canadian Groups Want Competition Review.* Retrieved from https://www.naturalblaze.com/2016/09/bayer-monsanto-acquisition-anti-competition-fears-competition-bureau-review.html

Top 10 Best Food Processing Companies in The World. (n.d.) Retrieved on August 4, 2009, from http://www.trendingtopmost.com/worlds-popular-list-top-10/2017-2018-2019-2020-2021/foods/best-food-processing-companies-world-largest/

Chapter 6
Aware Consumers Are Active Citizens Too:
Participation as a Possible Strong Individual and Collective Solution

ABSTRACT

Learning to pick up tools and data and use them according to our real needs, aware that we are all producers of information, it is the basic step to an active citizenship in the global world. From the example of the children setting up their "Virtual Museum of Small animals", from the school yard to the Internet, the author argues about how also average people, thanks to the spread of technology, can develop shared models of production in the cultural and more in general third sector of the economy. But important innovations and changes of perspective can be applied also to agriculture and industry, above all sharing more and more the knowledge, that is maybe a possible response to many problems of the world today, for a sustainable development in the planet and the peace among the groups and nations.

DOI: 10.4018/978-1-5225-3770-0.ch006

INTRODUCTION

Hardware, Software, Attitude

We can think we have to undergo the future, but we can also approach it as active actors. And something we have to change in our way of thinking, not only adapting to what goes on the market.

It is mostly a matter of attitude; and it is extremely difficult, for passive, TV addicted, individualistic people, to imagine a new, active role in picking up resources and using them, together with the others, sharing duties and pleasures, inside a superior organization which can be vertical, "company based", as traditionally, but also horizontal, collectively based, managed directly by the communities, locally and globally, if only adequate ways of management are found and properly worked out, for an actual impact on the real world.

Hardware are objects which can be taken, touched, above all watched, when they are so beautiful and stylish as the latest smart phones, tablets, laptops and "all in one" PCs are! Objects are produced by producers and bought by consumers, without any doubt or confusion, in the long wake of the industrial tradition (that is almost as old as the idea of free economy, how much: two centuries and a half?).

Lately, the trend of some sellers seems to be to keep far away from the consumers' minds any temptation of being themselves somehow producers of objects. So, for example, optical disc drives are cut off from the shining, highly refined and extremely *thin* "all in one" (all?) PCs Apple *iMacs* and Microsoft *Surface Studios* that we proudly exhibit on our desks. I go on thinking that it is hard to find even only one technical or practical reason for which a desktop PC should be *thin* but, as many signs suggest that is happening, more than for a working tool, many conceited protagonists of present "digital age" are probably searching, again, for the finest Narcissus pool in which they can mirror!

Is this the "future"?

Software way to things is maybe a bit more present-day. We have some manufactured goods whose way of working is not inbuilt, but depends on instructions given, in the end, by the humans, us included. So, though an "industrial" base does exist – a car and a sewing machine are designed for different duties – the chance in the future is to have machines matching more and more different approaches and styles, as real people are all different.

That is another scene, indeed, from the crowd of nerds gathering in a hall, all with the same brand laptop in their hands, as in a race of conformism and approval.

PICKING UP TOOLS, DATA AND A DIFFERENT STARTING POINT

The ability of picking around hardware and software, actually good for our jobs and intentions, as well as picking up just the data and the information we need, in the ocean of data and information running every day across the web, and giving them a sense according to whatever activity we are carrying out, it is the key to enter as aware actors into the present world.

The awareness that "machines" are not to compel us to change our lives according to their way of working, but are waiting from our instructions and orders, waiting to be discovered just by us for an infinite number of possibilities which are not in the instructions for use and maybe not even in the projects of their original creators, this can give to everyone of us another idea and attitude about the "future". And the matter is just to decide *who* gives instructions to machines, who sets up the future.

With other kinds of "technologies" and tools, we already do this when we use words to write always new, unpredictable texts, use notes to create never heard music and songs, use shoes for walking unexplored paths and pans for cooking dishes never tasted before. The time has come not to ask anymore the mechanical where we should go with our car!

But to understand what actually means to have, anyone of us, an incredible amount of possibilities of doing, expressing, creating, communicating, further multiplied by sharing everything with potentially thousands or millions of other people, it is difficult and likely to give dizzy, if we cannot see all this exemplified in simple, lived, intelligible works or activities.

I am presently engaged, since some years, in a project with children as "natural" as simple, as probably giving an idea of what I mean. It is usually taken by schools as a project of "science", but it is at the same time a project of video, photography, art, storytelling, social networking, web mastering, intercultural dissemination, environmental education… and what else? But for the kids, above all, it is interesting, amusing and exciting. They love it.

Writing of *The Children's Virtual Museum of Small Animals* (Beneventi, 2012), I have to tell what we and the kids do, which machines we use, which topics in theory and in practice we have to know and study, how we stay on the Internet, and many other things. But what we could consider well, from a theoretic point of view, are some connections and implications of a way of working like this, which until some years ago was simply impossible and so probably is quite innovative, but does not cause any of the problems usually related to the "innovation" in schools.

What have we so radically changed, in fact, but so naturally that even we do not realize it?

What is the difference from many situations in which "technology" is almost forcefully inserted into contexts and activities that have a hard to accept it?

Can we think – assuming a way of thinking like this – of applying the same or a similar pattern to other projects and activities not only in "science", but in art, information, social services, agriculture and industry?

Some years ago, at the "Global Junior Challenge" in Rome, I met people from a group working now with us, from Taiwan, Malaysia, Colombia and other countries in the world, who took part in the event and won a prize with their *Organic Garden Dream*, a project of biological gardening shared among the schools all over the planet. The topic is back on what we have already written about food and culture. We do no not have indeed the solution for every problem, but perhaps we have the opportunity, each of us, today, to be somehow the protagonist in "feeding the multitude", as never common people in history has been.

Though we have probably never thought of it before, with the right technological tools and attitude, working together we can.

FROM THE YARD TO THE INTERNET: AN EXAMPLE?

It is based on a very simple activity, almost "natural", having at disposal powerful, easy and cheap tools to take very close still and moving pictures, as compact video and photo cameras are. I refer to the *compact* devices, as to use a reflex camera, eventually with special macro lens, is not exactly "natural", nor essential, but an advanced choice for a better quality, implying not necessarily a professional, but a superior expertise.

Here we tell instead, as a starting point, of average kids of the primary schools or even kindergarten who, together with their average teachers, can go out in their school garden or yards and, just getting with their camera as close as possible to insects and spiders, take some pictures. They need basically only to be careful in approaching the animals, hope that they do not run or fly away, not move when shooting, as many times (not all) autofocus systems work and, if the light is good and with a bit of luck, some amazing and surprising pictures will be taken.

With recent sensors provided in cameras with more than ten megapixels, for the photographs of large and medium size subjects it is not even necessary to stay just very near to them, as it is possible to cut the picture at a later time and keep only the part we need, usually the central one, having anyhow an image large enough to be seen and magnified as we like. But this also gives chances to get pictures good enough from very small subjects, whose real appearance in details we hardly can imagine before watching in macro pictures. We photograph them as we can (that little "point", that "fleck") and we hope later with magnification we'll be able to see something more, as the shape and color of the body, a head, eyes, eventually wings…

To learn how to cut a picture, by the way, with any photo software on a computer, for those many people, teachers included, who are not able to do yet, does not take more than ten minutes!

The best results in capturing are achieved with the so called "super macro" lenses, allowing anyone to capture from 1 or 2 centimeters (instead of the 10-15 of normal macros position on compact devices or mobile phones) and available also on cheap machines. In this respect, it should be useful to know that the lenses of traditional video cameras (never really understood by average consumers and today limited in a small niche of the market) normally allow shooting just "stuck" to the subject, taking also still pictures.

But in the end, what is important is that during such an activity we get a number of good pictures (eventually chosen among many wrong) which we can watch, magnify, study, share, of insects, spiders, millipedes, as probably we had never seen, made just by ourselves. Every group, almost always, in any area with a little green, makes some interesting meeting and collects some good images.

No special preparation is needed at the beginning, no study on science of biology before, but simply, after, possibly with the assistance of someone already a little skilled, to start from our own pictures and try to understand a little more every day, knowing that to correctly recognize families and species is often not easy and that we have not to be hurry: shapes, colors, drawings; antennas, wings, legs; what is common in butterflies, bugs, beetles, comparing our pictures with those we can find in books and on the web, paying a great attention to details - that is what in normal life we seldom are used to do - and remembering that the references on line, for many reason, can be wrong or inaccurate, so that we have to find quite large a feedback, before accepting them. We could also, at a certain point, get in touch with some true experts, looking for them in museums, libraries, associations, universities, groups and forums on line, as many people probably are ready to give us a hand.

If this scene may look a bit complicated to be managed, referred to a single group or situation, we have to think it multiplied for many, linked one another, so that every picture, observation, research, connection got from one, it is as it was for all. So, when we need a picture, we can pick up it not in generic and impersonal web pages, but from a group of correspondents we are in contact with; when we need an expert, we have some of them already selected and connected with our own project.

In the end:

The Children's Virtual Museum of Small Animals is a website where multimedia documents are collected, based on the real experience of groups of children from many parts of the world. There, people can find photos and videos of insects and spiders, with scientific names and classification, place and date of discovery, and age of the class, group, or single kid who found it. There are also drawings, texts, and other things related to the real, possible, or fantastic meeting between children and small animals: voices from actual experiences, slide shows about "watching details," pictures of creations by artists close to kids' imagination, suggestions on how to use technical tools. Children there can act as protagonists in producing and sharing information, just like usually scientists, journalists, photographers, and video makers do, through the global information society. (Beneventi, 2012)

We are the "scientists", the discoverers, the photographs and the video editors; we are the fitters of the museum, of the web pages where we can meet and share information; we are eventually the publishers of books, videos and multimedia.

We have to count of course, especially at the beginning, on some professionals, on some "ready" virtual environments, as the pages and the groups on Facebook, the channels on YouTube and Vimeo, but after setting up the right technical organization, the Museum can also live as a standing alone enterprise, whose producers are kids, schools, groups and common people from all the world.

Cooperating and sharing in this way, everyone can take an active and significant part in a number of important realities, and global information in the world begins to be something really different, with many and good middle quality productions which begin to fill the void between those of the great companies and the familiar ones.

A SHORT CLASSIFICATION OF LIVING BEINGS, BEYOND THE BARRIERS OF SPECIALIZATION

This is, in my opinion, a good example of how, thanks to a proper use of technology and the sharing of information, as it is possible today and not in other times of the history, some fields of the human knowledge and production until now considered "specialized" are successfully practicable by all. As the authoritativeness of the professionals in that field comes from their superior ability of managing a matter shared with others, and not from the power of keeping the others completely out of that field, without possibilities of saying a word or doing something. Beyond a world unnecessarily divided into sectors and specific roles even at minimum levels; towards a general sharing of the basic knowledge and the enhancement of the skills of the specialists at the highest levels. That could mean a great global improvement in culture, in production, in quality of life, overcoming the huge waste of resources and opportunities caused by the general competition and exclusion as a rule in human relationships:

If technology is seen not only as a collection of complex gadgets and machinery, but also as the methods and techniques developed in the communities of practice that use these technologies, then we see that technologies create particular kinds of social spaces within which certain activities are more or less possible and more or less likely. (Jordan, 2014)

So, we start here from our observations of the pictures of the insects. We begin to look up on books and on the Internet and also, obviously, we ask to the experts. Though, we do not take always every word or piece of news as the absolute truth, as also the experts sometimes, due to their professional point of view, give answers right in general but not good for all our questions, as it is testified by the general difficulty for common people in understanding well what they write in books and web pages (in our case) on insects.

An example can be useful, very easy:

Kids are fascinated to know that some insects with black and yellow stripes are not bees or wasps, but "disguised" flies. But, how can we distinguish them from real bees and wasps?

An expert once told to me: "You can count the wings!"

What? Are you kidding me? I know that Hymenoptera have four wings and Diptera only two, but how can I count them? Wings are usually transparent, they often appear overlapped and even in very close macro pictures it can be difficult to distinguish them clearly. It is easier to see the two halteres which in flies are instead of the second pair of wings, but only watching much magnified photos and not in all. I can say this for sure, even if I am not an entomologist, because I have taken many macro pictures of flies.

So, again, how can I distinguish them possibly at a glance?

There are many other differences in fact, all watchable from the observation of the photographs, in the eyes and the legs for instance, but one is indeed very visible, even to the naked eye: the shape of antennas. Those of bees are not very but quite long, and articulated, as they use them more or less like arms in their seeking and rummage into the flowers, we have seen this in our pictures (wasps have not articulated antennas, as they do not rummage!). And those of flies, on the contrary, are very short and round. This is from our direct and green observation, but it is confirmed by the scientific name of the suborder in which flies are classified: *Brachycera*. That is from the ancient Greek:

brachys = short + *kera* = antenna

So, starting from scratch and only watching some pictures, any group of kids or common people easily can come exactly to the same conclusions of the scientists: amazing!

And at this point the experts are not the owners of a knowledge we cannot approach if not from them, but the ones who can help us to put together well or better the scattered pieces of a knowledge we are collecting in a significant part from our experience, direct or shared with others like us. And some of our observations are probably useful for the experts themselves, as we give them the occasion of looking at their matter from another point of view than the usual of their group and colleagues: good tips for the popularization of science, for which they can have an actual feedback.

Another short example, in the same direction, is from the words of six years old kids, looking at the projected magnified pictures taken with them and others. At a certain point they say, without any explanation or suggestion by the adults: “Bees suck and wasps cut” It is an empiric observation, as they have just seen the tongues of bees and the jaws of wasp, but it is also the main distinction made by the scientists between the two super families *Apoidea* and *Vespoidea*.

I remember that sometime later I was following a so called *sand wasp* browsing on flowers, and with my camera I captured it in good pictures with its long tongue into the flowers. But… if the kids said “Bees suck” they can’t be true wasps!

Popular names follow generally other principles than the scientific taxonomy, in this case the exteriors aspect of these thread-waisted insects, right called *wasps*. But if we go and see on the scientific side, we know that they belong not to the super family of wasps, but to that of bees: *Apoidea*.

In the end, we have collected these and other strong signs of distinction, according to the correct classification but also to what kids and “normal” people can see and notice (without deep researches, investigations on dead specimens, microscopic observations of embryos in a laboratory, as the scientists do): the short antennas of flies, the segmented ones of bugs, the quite short of grasshoppers and the very long of crickets; the different habit of spiders waiting on their webs, hanging from a thread or running after their prey as wolf spiders do; the slower walking of millipedes and the very quick of centipedes; the similarities and differences between snails and slugs; the difficulty of taking pictures of lizards, except when some kids are good in catching them… And from all this, as well as some additional readings we try, with the help of kids from the second grade (“do you understand?”), to obtain a summary in few, easy concepts, as simple as correct from a scientific point of view, of how the animal living beings can be divided.

So, there are two great categories of animals: those very simple, which do not have organs (sponges), and those more complex, with organs. Right?

Animal with organs are divided into two great categories: those that have a spiral structure (jellyfish and corals) and those with a left and a right part in their body. Is it clear?

The animals with a left and a right part are divided into three great categories: those without a skeleton (snails and earthworms), those with a skeleton outside (insects, spiders, centipedes) and those with a skeleton inside (as for example also we humans are).

Few strong ideas, clear and exact, and we all, kids included can now begin to classify any living being into a scheme that allows us to begin to understand.

THE EGGS OF THE EARWIG AND THE WORK IN THE INFORMATION AGE

The kids have lifted a stone and an earwig was under it. All the kids of second grade gather and watch. There is also a little heap of very small white balls and the insect now is stirring up and down around them. It takes one with its mouth, brings it into a hole, then comes back for another, but it seems to be insecure...

"Its eggs, they are its egg!" The children cry. We have just exposed a place where many small animals feel safe (under a stone) and now a mom earwigs is in a panic. While aside a woodlouse is walking, maybe a bit annoyed, but without hurry.

Here the scene is described in words, as it is natural in a book, but its memory is stored in a more "objective" document, a video we took in their occasion, with the right sequence of the events, times, words, the emotion in the voices of the kids, the "confusion" of the female earwig. That it was a female specimen, we can see from its straight pincers (whose scientific name is *cerci*, as male *Dermaptera* have curved ones) and we now also know that it takes care of its eggs.

This is first-hand knowledge, from direct experience, as scientists do. And from a quick research on line we can easily know that a scene like this has already been observed and we all agree, professional entomologists and kids:

Earwigs are among the few non-social insect species that show maternal care. The mother will pay close attention to the needs of her eggs, such as warmth and protection. . . . The mother will also faithfully defend the eggs from predators, not leaving them to eat unless the clutch goes bad. (Earwig, 2017)

More, reading around here and there, we also know, from the observations of the scientists of nature, that "earwigs raised without parents demonstrate limited maternal care of their own offspring" (Earwigs raised without parents demonstrate limited maternal care of their own offspring, 2015). So, love for them – just as it is for humans – comes from the experience of love. We tell this to the kids and their emotional engagement in all the story grows.

Insects and other weird creatures have not necessarily to be represented in human shapes, with a face like ours and the same kind of smile and eyes. They can be well understood and liked by kids just as they are, with their "nose" on antennas (as flies) and their "ears" on legs (as grasshoppers). So they are even more interesting, for curiosity and wonder is the light, amazing side of diversity (as ignorance and fear are its dark side).

But knowing for instance that earwig is a good mother with her children, it can be very important in order to feel those creatures not only so different but also so close to us, to the children themselves, and not for the exterior appearance, but for the basic facts of life. Earwig moms love their children.

Looking around, young earwigs are not difficult to be found, with their soft back and pincers, more gentle and delicate then the adults, as human kids are.

At this point, the reader may ask a couple of questions. The first is probably where I am going to with all these lines and paragraphs on insects. The second is the objection that, obviously, not every group of kids has always the possibility to come across a scene like the one have described, with the earwig and the eggs, as such a meeting is rare, and so the quality of the experience depends in the end from luck.

But for kids used to enrich their searches and observations on the field with close macro photographs and video shots and to share them with others, we can probably say that almost every pictures taken by any other kid of the world, or even by the adults, including the professionals, it works as if it were have been taken by the kids themselves. This is right especially for videos, in which they can ear and feel the voices and the emotion of other kids, across the different nations and languages.

So, if the information of the earwig mom and its eggs comes directly from video shots just taken by or with that group of children, this is a great thing for them, and the emotional engagement is deep. But it happens also with other kids, who have met other insects (flies disguised as wasps, "houses" or "streets" of ants, bugs walking or resting on leaves). They naturally "add" the new emotion as an extension of their own, as a part of the same story. And also any information coming from the adult photographers or video makers is then taken maybe not with the same emotion, but as the natural aid or contribution that adults are expected to give to the children, when working together or somehow in contact on the same subject.

So any group, in any part of the world, who has taken photos or videos of insects and spiders in the garden or yard of their school, any group has almost the same experience of the eggs of the earwig.

This is magic, this is the huge potentiality of sharing, as nowadays video devices and web connections make it possible, easy and natural.

This is the magic of how kids and creative humans can stand in front of the works and knowledge: the great ability, having done something or begun to do something, to *feel* what others do as if it were their own. For example, if young children give their kindergarten teacher cardstocks eventually colored and drawn by themselves and then the teacher assembles the pieces in puppets, the children say: "*We* have made puppets!" and this is not to take for themselves the merit of someone else, but the actual reality, as they have been an active part in the work.

That is also (it could be) the "magic" of a way of working just typical of the information age, when the model of the self-made individuals always competing one against the other is put beside or possibly replaced by another model (Snow, 2015?), in which the works of many naturally converge into a final result which is the fruit of the collaboration, direct or indirect, intentional or unintentional of a lot of people. This is, if we think, just what happens with most of the ideas circulating today, despite the attempts of owning or copyrighting them.

Such point of view, far from denying the merits of the individuals, can be in the end more effective and satisfying, as well as relaxing for all, and even profitable. It is the way of working that applies in the development of free software, in Creative Commons licenses, the way of working in fact applied (not always with intention, as actually many "stole" form the others) in the first years of personal computers and with much more intention in the setting up of a great part of the World Wide Web and other collective technological creations of the present time. But it has also always been, along the history, the point of view of the communities of scientists:

The hacker ethic of Richard Stallman, father of the GNU project (Copyleft: all rights reversed) does not oppose making money, just making money by closing off information from others. He is proposing a new kind of free-market economy: a free market economy in a much deeper sense than in the normal capitalist vocabulary, but still a capitalist economy. . . . The scientific ethic entails a model in which theories are developed collectively and their flaws are perceived and gradually removed by means of criticism provided by the entire scientific community. Of course, scientists, too, have chosen this model not only for ethical reasons but also because it has proved to be the most successful way of creating scientific knowledge. All of our understanding of nature is based on this academic or scientific model. (Himanem, 2001)

SHARING PRODUCTION: MAPS AND NEW POSSIBLE POST-INDUSTRIAL HORIZONS

I was recently watching a TV series of not many years ago, when already personal computers and Internet were a common experience for many. For a crime investigation, they used satellite photos and could get them because they were FBI agents…

Maybe we should think more of some incredible possibilities which are now part of the daily life of all, as for example looking up the maps of all the countries of the world, in many cases so close and detailed that we hardly believe not to see just exactly from above the same cars in the parking next door we have just met a few minutes before, walking outside. Even more, we can even "land" and go along the roads, watching houses and shops, turning left or right, "walking" to and fro.

I do not know if is a positive thing that all this may appear natural to a "digital native", as the work behind is really huge. Not only an incredible number of photographs from the sky, put together with a sophisticated software which makes available to anyone all the earth from above in few clicks and zooms in and out, but the incredible experience at ground level, now possible by the capillary work made in every part of the world for Google Street View and other similar applications: "Most photography is done by car, but some is done by trekker, tricycle, walking, boat, snowmobile, and underwater apparatus" (Google Street View, 2017).

Thinking well of all this, we can have an idea of which never even imagined opportunities we have today, in the post-industrial world. But to approach them from the right point of view is not quite easy neither for the old generations, who hardly follow the times of technological innovation, nor for the "natives", who do not know the story and the differences from the past and are used to simply take already made solutions as pre-cooked food from the fridge or the microwave oven.

A comparison can be useful with another very "natural" thing that we all use a lot, every day, all day long, without asking anything about it: electricity.

Behind it, normally, there is a huge industrial as well as artisanal organization, from power plants and lines to the cables and electrical outlets in every house, office or building. We all are essentially *consumers* of electricity, though in latest years the spread of photovoltaic technology is giving also to average people the opportunity of being somehow producers and dispensers of small amounts of energy, which could be probably increased in the future, depending on the progress of technology and on the political choices of the humans. Anyhow, we do not take part with our direct activity in *making* electricity, as solar panels on our roof are to be bought from their manufacturer and are to be installed by specialists.

Electricity is a very typical industrial produce, mainly based on an extremely complex network which need a centralized control and maintenance, but which now, thanks to the technological progress, can be in part redesigned giving the possibility of choosing, for instance, if to have a unique big power plant producing one million megawatts or one million small plants producing each one megawatt, and also of having many self-producing unities without a connection to the network, as they are already solar power led lighting points and they could be for instance in the future electric cars with a photovoltaic nanotechnological paint taking more energy directly from the sun and less from a battery.

Maps, on the other side, are a very typical post-industrial produce. They are based, of course, on a satellite network, and often also on the Internet (as many map apps do not work without a web connection), which are complex and big industrial infrastructures too, but their construction, in the same way of web pages and all web based activities, could be shared and basic much more than most of us do not dare to imagine.

Having the right software (it could be also free software, developed by a large community, as Linux, Mozilla, Libre Office, the GIMP), and having the access to a satellite network (paying for the service), nothing inhibits to any group of people to take their own street photographs (by car, trekker, tricycle, walking, boat, snowmobile, and underwater apparatus, any of them available also to common people today) and even aerial photos, using drones, and to develop its own map.

Not necessarily this is in competition with what big international companies have done as Google, Nokia (Here Maps), Microsoft (Bing Maps), or Yandex (popular in eastern Europe), having means, money, time and know how much vastly superior, as to make good maps is not easy indeed - as the experience of Apple in 2012 testifies, when the first attempt to replace Google Maps on the iPhone with their own system was a failure - but there also other possible options: maps of small territories "updated" for special events or purpose, eventually integrated in the maps of the big companies; "didactical" maps made in part by students, testing the right techniques of taking pictures of a land; "fantastic" maps for creative works, in which the territory is taken and transformed, as it happens in some video games; participation of local communities in the updating of the big maps themselves, having got the right instructions; and so on. It may be complicated at a first glance, but it is not impossible at all, as a centralized and proprietary management of all the steps of the work is not actually necessary and the technical means are in fact spread and available among common people.

So, we find that *Mapillary*, for example, is a community of "people working together to photo map the world" (Mapillary, n.d.), and it is interesting and somehow surprising to read the words of some members:

It should be crowd thing, a human thing, not a Google-only thing.

I mapped my home town with an old bike and a home made mount.

Why rely on Googles car when you can do it yourself?

There are already in fact also many examples of applications based on maps, from every part of the world, where points and paths are indicated of solidarity consumption, special artisanal productions, alternative accommodation; there are many services, not only commercial, using different forms of *geolocation*, whose basic tools are in the hands of everybody, such as mobile phones, or in the possibility of all, such as web cams and info points based on QR codes.

Many people here and there in the earth are imagining incredible things starting from the huge number of possibilities we have to manage, process, transform virtual worlds created and directly handled by anyone of us around the real world we live in, possibly also changing a part of our behaviors and habits in the real world. Mainly they operate above all in the field of vanguards, of science, of art.

So writes Carlo Infante on his site "Urban Experience":

Walkabout is a format of "performing media" where peripatetic conversations combine with nomadic radio transmissions, for a participatory exploration. . . . The term was coined by Australian white landowners to refer to Aborigines who disappeared from their property, and referred to as "gone walkabout". . . . Urban Experience plays with this definition by associating it with "talkabout", thus relaunching urban explorations which combine simple things such as walks and conversations with the unprecedented complexities of the Performing Media-storytelling in which the participatory narrative is inscribed in the action, "augmented" by the use of radio and web media. (Open House Roma, n.d., loosely translated by Paolo Beneventi)

And Jeff Paul:

Digital banality, as used here, refers to the unoriginal and uninteresting excesses of the Internet. Our relentless clicking and scrolling, in search of "content" confronts the idea of banality on multiple fronts. Shulgin's Form Art, in removing all content, forces users to focus on the structures of their digital environment while still performing clicking and scrolling actions. These truly banal HTML elements now become original, interesting, and entirely revealing. (Paul, 2017)

Such works and projects, however, hardly affect the great mass of consumers, who simply *buy* devices and apps and do not even dream they could be active protagonists in a movement.

The word "map", when speaking of digital applications, refers not only to the navigation and driving software which help us to move in the real geography of the world, or to training programs which record our bike or running tours, but also to the so called "concept maps", that are available in very interesting software in latest years.

To draw basic concept maps a PC is actually not necessary, and many, at work and in education, have been used to draw them on paper, helping the representation and memorization of many subjects:

According to the "dual-coding" theory of information storage (Paivio, 1991), information is processed and stored in memory in two forms: a linguistic form (words or statements) and a nonlinguistic, visual form (mental pictures or physical sensations). The way knowledge is coded in the brain has significant implications for teaching and particularly for the way we help students acquire and retain knowledge. . . . Yet, it is well established that showing children how to represent information using the imagery form not only stimulates but also increases activity in the brain. (Marzano, 1998) (Birbili, 2006)

But having now many and also sophisticated freeware and shareware (generally with a basic free version) programs available, the creation of maps with a computer is not only a cosmetic way for improving their presentations (as many unnecessary works in Power Point), but can be a very interesting occasion to find out our personal capacity of organization of tasks and arguments in forms of tables, having from different programs various and different solutions, possibilities and ideas for representing, ordering, linking interactively, and in the end have shown and understand better what we are working on, with a beautiful, impressive and amusing graphic (see Figure 1).

Figure 1. Example of graphics created using technology: This book at a glance

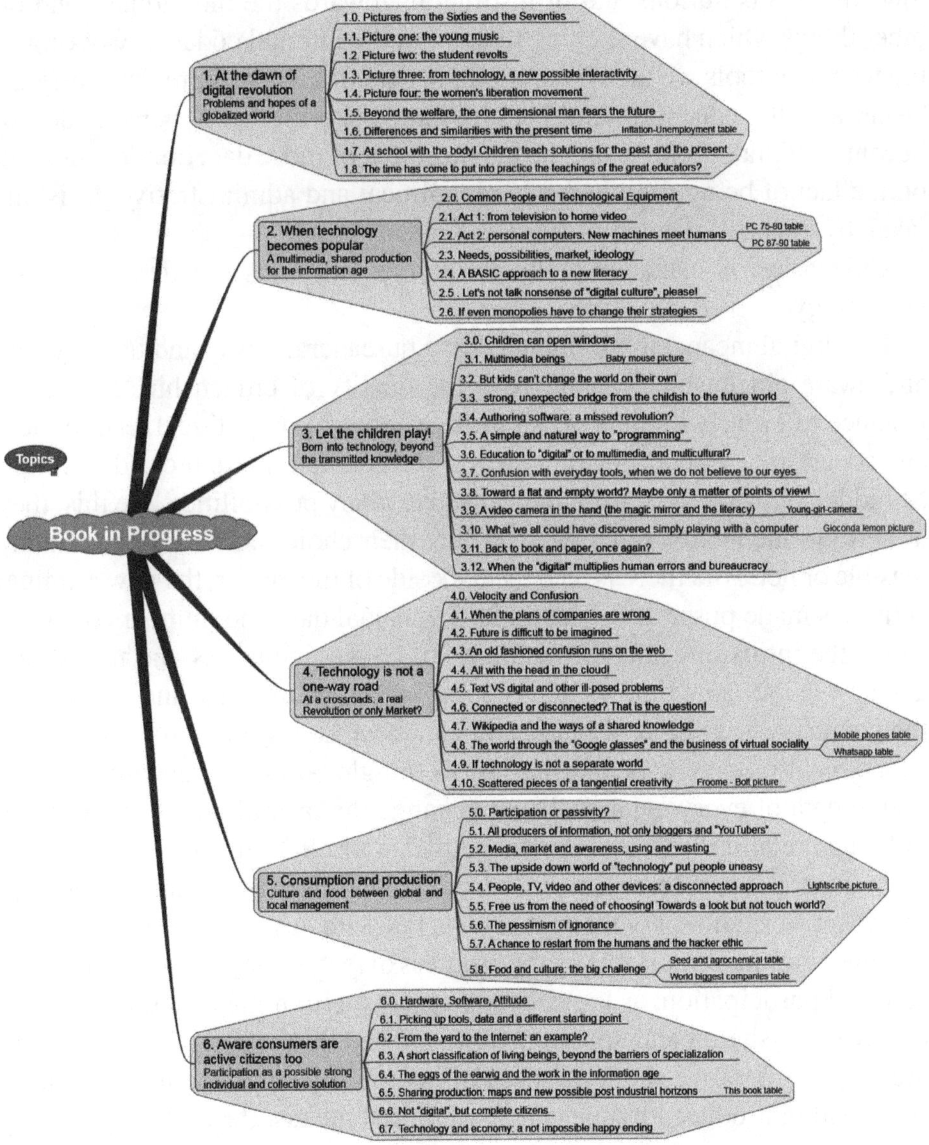

NOT "DIGITAL", BUT COMPLETE CITIZENS

Personally, I do not understand what many people mean when they speak of "digital" citizenship. Why using digital devices should change the quality of citizenship more than driving a car, riding a horse, eating spaghetti, wearing shoes or shaking hands? The quality of citizenship depends on the actual participation to the political and social life, on the communication among persons, on the respect for the environment, on the loyalty of individuals towards the institutions and of institutions towards the individuals, and on other things which have nothing to do with the internal codes of working of the technical tools we use. Even if all the data of citizens in registry offices, electoral rolls, education and health systems are digitized, persons do not become "digital" for that and the management of those data does not depend on the fact of being digitized, but on political and administrative decisions taken by men and women in flesh and blood.

Citizenship is a matter of human beings and it is not determined by technology.

But digital means, if not intended in a bureaucratic way and used by free and aware citizenship, can improve the quality of citizenship, giving for instance to everybody new possibilities of control on political, economical and social life, so that illegal practices and corruption are more difficult to be hidden. Obviously, if digital means give many possibilities like this, they are always men and women who, which their choices, make them actually feasible or not. And there is also the dark side of the matter, the new criminal activities made possible by the theft of personal data and numbers of credit cards, the intrusions into the data base of banks, industries and ministries, and new at the same time virtual and real scenarios of peace and war among the countries, new field of battle and conquest for legal and illegal powers, economical monopolists and mafias, as the global interconnections of the digital data of everyone actually can change the overall quality of all sorts of human relations in ways which are difficult even to imagine.

In general, apart from a direct participation in the life of a country or of the entire world not only through a vote, but with any other new instrument of democracy made possible by the progress of technology, also more indirect forms of participation, as the collective production of maps of our example, or any other shared and spread activity carried out in communities, surely are ways to improve the awareness of common people about the changes so recently introduced by the spread of digital means, the ability of dealing with them, and in the end the quality of the life of all the participants, as active citizens.

An important step towards active citizenship is also to invert the way we have accustomed to in recent years of expecting to understand and explain people, society and facts with titles, short definitions, *tweets*.

Any superficial simplification, if not supported by a deeper reflection, makes it increasingly difficult to manage the reality by the most, leaving the action always in the hands of someone else and foreclosing the perspectives of change.

A strong example – here already quoted several times - is the label that someone has given to the youngest generations as "digital natives". It comes from a shallow observation from outside of the ease with which even very young children juggle with gadgets such as mobile phones and tablets, assuming a series of skills that actually are only in the minds of the observers, often without any scientifically correct approach:

How students and teachers overcome barriers when promoting digital citizenship was not found in the literature. How the tenants of digital citizenship can be used to promote ethical and moral use of technology in networked and global communications was not found in the literature. The perceptions held by middle school teachers' and project administrators on and familiarity with digital citizenship as well as their challenges and successes associated with integrating digital citizenship into the curriculum when combined with global collaboration and social media are missing from the literature. Research is needed to provide answers to those questions so that students can learn about the use of technologies within a global arena. My study was needed to analyze strategies that promote the ethical and moral use of technology among middle school students as they become productive members of a technology - infused global society. (Snyder, 2016)

All another perspective – other example - is to speak of philosophy with children. It is not in this case teaching a subject, but to be curious and really interested in what kids may think about the great issues of the world, assuming their point of view as important, after all we have already seen here about the possible central role of the youngest in overcoming some of the great problems of present times, quoting *animazione teatrale*, the teaching of great educators, the possibility or not we all have to "enter the kingdom of heaven".

During an experience lead by Luca Mori, a researcher of the university of Pisa, ten years old kids were requested to comment a passage from Sophocles' *Antigone*:

Numberless are the world's wonders, but none

More wonderful than man; the storm-gray sea

Yields to his prows, the huge crests bear him high

Earth, holy and inexhaustible, is graven

With shining furrows where his plows have gone

Year after year, the timeless labor of stallions. (Sophocles, 278-284)

How much could interest an old tragedy to kids used to living with video games? Much, in fact. They express their ideas:

In my opinion, the human being is both wonderful and terrible. Wonderful in the sense that he invented many things, but doing so he polluted the world and also destroyed animal habitats, the green areas.

He wants to mean that man is both constructive than destructive.

I think that man is a part of what we can say wonderful. As it has been said, he can destroy the environment around him without worrying about the damage he can cause to himself and to other species, to other living beings who however have a life and need to ... be able to build their own life without destroying what he has already created. . . . But it is also wonderful as he found out his DNA. . . . The man was also very smart to discover the DNA because he discovered something very important for mankind. I also think that man is wonderful, but for him to be the most shining he destroys things, the wonderfulness of animals, only for him to prevail over everything. (Mori, 2014, loosely translated by Paolo Beneventi)

Going and searching about in the world we find - luckily – that this idea of talking of philosophy with kids is not so weird, as for example: "Childhood & philosophy is a three-time-a-year journal dedicated to the intersections and interfaces between philosophy, childhood, children's philosophies, and philosophical inquiry with children. It is a result of a co-working between NEFI (núcleo de estudos de filosofias e infâncias), which is part of the post-graduate program in education of Rio de Janeiro's state university, and ICPIC (international council for philosophical inquiry with children). It publishes regularly three electronic volumes a year. At least one of these is printed". (website)

In times of collective short memory, when everything in few months is told to become "obsolete", it is probably very important to stress, with true examples, the possible connections between the thought and the problems of nowadays citizens of the earth and those of their ancestors lived hundreds and even thousands of years ago. To feel part of the history is something that helps us to live, as well as to understand better from where some ideas come, which we maybe believe we are inventing now from scratch, as well as some problems, that have not always to be faced if they were totally new. It is also important to overcome the used mindsets, which in changed situations risk being counterproductive:

The ethos of the original academic and the hacker model - well summed up by Plato's idea that no free person should learn anything like a slave - is totally different from that of the monastery (school), the spirit of which was summed up by Benedict's monastic rule: "It belongeth to the master to speak and to teach; it becometh the disciple to be silent and to listen." The irony is that currently the academy tends to model its learning structure on the monastic sender-receiver model. The irony is usually only amplified when the academy starts to build a "virtual university": the result is a computerized monastery school. (Himanem, 2001)

TECHNOLOGY AND ECONOMY: A NOT IMPOSSIBLE HAPPY ENDING

Steve Wozniak, co-founder of Apple, "summed up the elements that motivate his action in the speech he gave upon graduating from the University of California at Berkeley in 1986: "You don't do anything in life unless it is for happiness" (Himanem, 2001).

Apparently, in latest years, it is not in the direction of happiness that the world is going to. Also without considering hunger and lack of water in many parts of the planet, as well as wars and terrorism, "normal" production of goods, widely displaced in poorer countries, means in general worse working conditions in factories than before. And many complain the dark side of new technological jobs, far away from the glamour of billionaire companies:

Most Internet workers experience high levels of risk and insecurity in their work: they are forced to responsible for their life after retirement, which was previously the responsibility of the state; some are pushed to experience karoshi (death from exhaustion) due to the high financial pressure in their work and lives. (Fuchs & Fisher, 2015)

In general, and according to the world we know, taking part in social life is not directly associated to happiness, above all as it means also responsibility, which is what many people indeed tend to keep far. And indeed, in a competitive and individualist context, the idea of responsibility is heavy. But also it could can be very lighter if we get used to think it *shared*, that is other from the load on the shoulders of someone else of all the weight of commitments and annoyances.

In our still old industrial and "pre-digital" social and economic organization, and used as we are to look at the objects and appearance more than at the relations among people and their actual quality of life, we can hardly imagine which could be the opportunities of having an increasing part of the production based on shared models rather than on competitive ones, deeply and widely exploiting the possibilities given by a technology knowingly chosen and selected according to our needs and intentions.

Production in information and services, as in the example of maps, may benefit from an increasing level of participation by many common people, moving the job of the professionals at the superior stages, as collectors, organizers and finishers of widespread activities, possibly very limited and easy, carried on also on a voluntary base, or in leisure time. Something like what we already do usually, without any impact on production, when we talk on something in a social network, or we do with little impact tagging with stars a software or a device in on line shops, a hotel or a restaurant in *Trip Advisor*, and so on. Something like what we do often unaware when, in our browsing the web, more or less legally we are tracked about our movements, researches, preferences and tastes, so that the right ads can appear on our Facebook page (Hill, 2015).

And if we were asked for whom we'd like to be tracked? If instead of a commercial operation behind our backs, it could be a conscious participation in something we know and can follow? We could have less sales platforms and more sharing platforms, without excluding sales, but putting them on a higher level.

We hardly can imagine, if the new technological means were not only used mainly in that old perspective of selling and buying goods, but really inventing new fields for the human mutual activities… This could be, diffused and properly organized, the new "voluntary work" in the information age, not only a way to spare on true works, with someone doing for free what is not done, though useful and sometimes essential, by public and private providers of services. And maybe, so improving the quality, satisfying needs and generating wealth, it could be also a way for finding resources to have the real jobs paid, in the welfare labor for instance and in general in the so called third sector. As we had also other resources available by lowering, at this point, the wages of many top managers, whose importance is reduced in a system of production based on shared activities.

A mass active use of hardware and software, in their possible interconnections between work and leisure time, producers and consumers, the top and the base, it could help perhaps, more than only – as always many repeat as a mantra – to boost consumption, also to redefine a significant part at least of the service sector of the economy.

We can take for good the definition of Simon of the functioning of the market:

Market processes commend themselves primarily because they avoid placing on a central planning mechanism a burden of calculation that such a mechanism, however well buttressed by the largest computers, could not sustain. Markets appear to conserve information and calculation by assigning decisions to actors who can make them on the basis of information that is available to them locally that is, without knowing much about the rest of the economy. (Simon, 1996, p. 34)

The central planning mechanism is not necessary to be intended as that of an old soviet government, but it can be referred also to the plans of oligopolistic companies trying to drive the future along a one way road, which are likely to fail when the actors on the real market, thanks to the capillary diffusion of technological means as good for production as well for consumption, according to their actions performed worldwide can actually choose other directions and other ways.

As it concerns the primary sector, if community-supported agriculture is surely not the final solution to feed all the Planet, it can give a contribution - and we do not know in what part, if we do not earnestly try to find out - practicing it and seeing how it can be integrate with the different kinds of traditional and industrial agriculture.

Often the "old fashion" view of many alternative farmers, distrustful and hostile to a technological progress in the hands of multinationals, has been a barrier to make us see the possible prospects for the future of this type of setting. But what does it really mean the possibility we have today, not only in theory, for the first time in the history, of connecting all together hundred and millions of small farmers all over the world, optimizing methods of work and costs of production, storage, shipping?

We can hardly imagine!

Not to overcome completely the great world companies, which at this point are called to an important job, but not exclusive, having as a limit – too much? Maybe the least in democracy! – to respect the life and the choices of local communities, this can be an alternative system of feeding a part of the Planet, more adequate for example to face the environmental and climate changes, on a regional scale:

A survey has been conducted to find out potential technology, related to the use of the Internet and mobile among farmers in the state of Punjab in India, to deliver agriculture-related information to them. Results show that agricultural information system needs to be developed based on the mass communication technology such as mobile systems. It is also noted that localization and native language of farmers are the concerns to be incorporated into the systems. It has been focused that the use of soft-computing techniques in conjunction with communication networks, for inferring the decision regarding best practices for agricultural activities, is helpful in the development of these systems. (Jain, Singla, & Kumar, 2014)

Using proper networks to monitor the situation, it can be also easily reduced the food waste and an answer can be done in time to the food shortages, anywhere. The logic behind, of course, is to be in this case to earn the right money from people's work more than only to make a profit anyhow, as technology is a great support, but does not decide instead of humans and, in the end, it is a matter of ethics and politics.

To find out if forms of sharing economy are applicable in a significant way to the industry too, it is not the main goal of this work. But the matter remains – to say it with Pekka Himanen again, not because he's my preferred guru, but as I have quoted him here many times and loaned some ideas of him, that is just *sharing*! – if the "virtual university" of the academy is necessarily to be computerized into a monastery school, that's the great contradiction of present days.

In latest decades, centralized multinationals have outsourced a great part of the manufacturing of any goods in countries where production relations are more favorable for profit, with the result that, if consumers can get cheap produces, jobs are lost in the mature capitalist countries and new forms of exploitation are established not only in developing countries, but also in mature ones, on the mass of immigrants and new unemployed, who have to accept conditions of work that before would have been unthinkable. Social troubles worldwide are the dark side of present "digital" era, in which the power just of digital means of eventually solving this problem is in fact canceled out by the habit (need, choice?) to sell them in the same way how pots and pans, toothbrushes and pieces of furniture are sold.

There is also a planet to be saved because, however we think about pollution and climate change, the impact of human activities on the earth is a problem which mankind has to face before it is too late.

It is possible, for instance, thanks to instant feedbacks, to develop new systems of relationships between producers and consumers, so that not only personalized goods are made according to requests (just that color for a car, or a sweater, as the customer likes), but also the amount of production fit more the actual demand and wastes and inventory excesses are reduced.

It is possible to adapt the diversification of products, which are usually replaced by others in few months without leaving the time to the consumers to really test them, to real feedbacks not only from the "global" market, but above all from the niche sectors.

It is possible to offset the decline in sales and earnings with a decline in investments and costs, to increase the average duration of the hardware, improving the performances much more via software. As the economical, social, political, ethic, psychological and even philosophical aspects of software, as the real engine for the future communities of humans (Accoto, 2017), are probably to be considered much more, not only in academic studies.

The idea of a possible auto regulated society based on free market and competition comes in fact from the 18th and the chronicles of capitalism tell of how in the world a general progress and increasing welfare have been achieved often through protectionism, colonialism, imperialism, world wars.

The perspective of a sharing economy is certainly much more related to what present times offer:

The speed of growth with which sharing systems have spread suggests that the sharing economy might represent a serious threat to established industries, due to fewer purchases and consequent distress in conventional markets. The Internet has made sharing enormously simpler. And for consumers, it seems to hold the potential to unite cost reduction, benefit augmentation, convenience and environmental consciousness in one mode of consumption. (Matzler, Veider, & Kathan, 2014)

Degrowth movement (Weiss and Cattaneo, 2017) - that is not recession, but a different idea of development - depending on points of view, but above all on its theoretical, practical and technological bases, can be a romantic dream, but also a earnest attempt to design a realistic future for the world economy, as many analysts even of sure capitalistic faith, are doubtful that, after the long world crisis begun in 2008, everything can start again exactly as before.

In general, active citizens of the global world have the opportunity today, technically speaking, to radically change the social, economical and political landscapes of the Planet Earth. This is possible thanks to the new technologies and how the planet is changing depends, in the end, on how the technologies are used, if we take the road of participating or delegating, of centralism or sharing,

Active citizens of the global world are aware that, living in an interconnected planet where the means are accessible to everyone, all of us have a part of responsibility in that big choice, for the present and the future.

REFERENCES

Beneventi, P. (2012). The Children's Virtual Museum of Small Animals: From the Schoolyard to the Internet. In *Didactic Strategies and Technologies for Education: Incorporating Advancements*. Hershey, PA: IGI Global.

Birbili, M. (2006). Mapping Knowledge: Concept Maps in Early Childhood Education. Early Childhood Research & Practice, 8(2).

Earwig. (2017). Retrieved on September 14 2017 from https://en.wikipedia.org/wiki/Earwig

Earwigs raised without parents demonstrate limited maternal care of their own offspring. (16 November 2015). Johannes Gutenberg University Mainz. Retrieved from http://www.uni-mainz.de/presse/19863_ENG_HTML.php

Fuchs, C., & Fisher, E. (2015). *Reconsidering Value and Labour in the Digital Age*. Basingstoke, UK: Palgrave Macmillan.

Google Street View. (2017). Retrieved on September 14, 2017 from https://en.wikipedia.org/wiki/Google_Street_View

Himanem, P. (2001). *The Hacker Ethic and the Spirit of the Information Age*. New York: Random House.

Jain, L., Singla, R. K., & Kumar, H. (2014). Assessing Mobile Technology Usage for Knowledge Dissemination among Farmers in Punjab. *Information Technology for Development*, *21*(4), 668–676. doi:10.1080/02681102.2013.874325

Jordan, B. (2014). Technology and Social Interaction: Notes on the Achievement of Authoritative Knowledge in Complex Setting. *Talent Development & Excellence.*, *6*, 1.

Mapillary.com. (n.d.). Retrieved from https://www.mapillary.com/

Matzler, K., Veider, V., & Kathan, W. (2014). Adapting to the Sharing Economy. *MIT Sloan Management Review*, *6*(2), 71–77.

Mori, L. (2014). *Antigone elementare: Filosofia con i bambini nella scuola primaria*. Pisa: Edizioni Mnemosyne.

Open House Roma. (n.d.). *Walkabout, esplorazioni partecipate: Viaggiamo dentro noi stessi quando ci ritroviamo in luoghi che ci ricordano cosa cerchiamo. Sd.* Retrieved from http://www.openhouseroma.org/2017/sito/walkabout-%E2%80%9Cflaminio-dal-maxxi-allex-teatro-politecnico%E2%80%9D.html

Simon, H. A. (1996). *The Sciences of the Artificial*. Cambridge, MA: MIT Press.

Snyder, S. (2016). *Teachers' perceptions of digital citizenship development in middle school students using social media and global collaborative projects*. Ann Arbor, MI, US: ProQuest Dissertations Publishing.

ADDITIONAL READING

Mori, L. (2017). *Utopie di Bambini, Il mondo rifatto dall'infanzia*. Pisa: Edizioni ETS.

Snow, C. C. (2015). Organizing in the Age of Competition, Cooperation, and Collaboration. *Journal of Leadership & Organizational Studies*, *22*(4), 433–442. doi:10.1177/1548051815585852

Weiss, M., & Cattaneo, C. (2017). Degrowth – Taking Stock and Reviewing an Emerging Academic Paradigm. *Ecological Economics*, *137*, 220–230. doi:10.1016/j.ecolecon.2017.01.014 PMID:28674463

Related Readings

To continue IGI Global's long-standing tradition of advancing innovation through emerging research, please find below a compiled list of recommended IGI Global book chapters and journal articles in the areas of activism, free speech, and protesting. These related readings will provide additional information and guidance to further enrich your knowledge and assist you with your own research.

Adewale, A. (2017). Does Bureaucracy Stifle Moral Agency? *International Journal of Public Administration in the Digital Age*, *4*(2), 1–13. doi:10.4018/IJPADA.2017040101

Adugu, E. (2014). Boycott and Buycott as Emerging Modes of Civic Engagement. *International Journal of Civic Engagement and Social Change*, *1*(3), 43–58. doi:10.4018/ijcesc.2014070104

Adugu, E. (2016). Political Consumption as Supplement to Conventional Political Participation in Promoting Social Change. *International Journal of Information Systems and Social Change*, *7*(1), 47–62. doi:10.4018/IJISSC.2016010103

Adugu, E., & Ampadu-Ameyaw, R. (2014). Attitudinal Measures of Political Consumption as a Form of Civic Engagement in a Developing Country. *International Journal of Civic Engagement and Social Change*, *1*(4), 28–46. doi:10.4018/IJCESC.2014100102

Al-Jarf, R. (2015). Discourse and Creativity Issues in EFL Creative Writing on Facebook. *International Journal of Signs and Semiotic Systems*, *4*(1), 54–81. doi:10.4018/IJSSS.2015010103

Al Khateeb, A. A. (2017). Designing a Wiki-Based Course for Enhancing the Practice of Writing Skills in the 21st Century: Moving from Theoretical Grounding into Practical Knowledge. In K. Bryant (Ed.), *Engaging 21st Century Writers with Social Media* (pp. 259–279). Hershey, PA: IGI Global. doi:10.4018/978-1-5225-0562-4.ch016

Anthony, S. M., & Zhang, W. (2017). Alternative Tweeting: A Comparison of Frames in Twitter's Political Discourse and Mainstream Newspaper Coverage of the Singapore General Election of 2011. In M. Adria & Y. Mao (Eds.), *Handbook of Research on Citizen Engagement and Public Participation in the Era of New Media* (pp. 324–343). Hershey, PA: IGI Global. doi:10.4018/978-1-5225-1081-9.ch018

Bacallao-Pino, L. M. (2016). Radical Political Communication and Social Media: The Case of the Mexican #YoSoy132. In T. Deželan & I. Vobič (Eds.), *R)evolutionizing Political Communication through Social Media* (pp. 56–74). Hershey, PA: IGI Global. doi:10.4018/978-1-4666-9879-6.ch004

Balint, B., & Gustafson, J. (2015). Nobody Knows You're a Dog but Everybody Knows You're a Republican: Finding Diversity in Political, Religious, and Personal Association on Facebook. *International Journal of Information Systems and Social Change*, *6*(4), 1–16. doi:10.4018/IJISSC.2015100101

Bobick, B. (2017). Promoting Civic Engagement through University Curricula. In L. Hersey & B. Bobick (Eds.), *Handbook of Research on the Facilitation of Civic Engagement through Community Art* (pp. 218–235). Hershey, PA: IGI Global. doi:10.4018/978-1-5225-1727-6.ch011

Braden, E. G. (2017). What Can I Do?: Using Critical Literacy and Multimodal Text Types to Enhance Students Meaning Making and Talk. In C. Martin & D. Polly (Eds.), *Handbook of Research on Teacher Education and Professional Development* (pp. 481–498). Hershey, PA: IGI Global. doi:10.4018/978-1-5225-1067-3.ch027

Bright, A., & Gambrell, J. (2017). Calling In, Not Calling Out: A Critical Race Framework for Nurturing Cross-Cultural Alliances in Teacher Candidates. In J. Keengwe (Ed.), *Handbook of Research on Promoting Cross-Cultural Competence and Social Justice in Teacher Education* (pp. 217–235). Hershey, PA: IGI Global. doi:10.4018/978-1-5225-0897-7.ch011

Bruner, M. S., Valine, K., & Ceja, B. (2016). Women Can't Win: Gender Irony and the E-Politics of The Biggest Loser. *International Journal of E-Politics, 7*(2), 16–36. doi:10.4018/IJEP.2016040102

Bryant, K. N. (2017). WordUp!: Student Responses to Social Media in the Technical Writing Classroom. In K. Bryant (Ed.), *Engaging 21st Century Writers with Social Media* (pp. 231–245). Hershey, PA: IGI Global. doi:10.4018/978-1-5225-0562-4.ch014

Buck, E. H. (2017). Slacktivism, Supervision, and #Selfies: Illuminating Social Media Composition through Reception Theory. In K. Bryant (Ed.), *Engaging 21st Century Writers with Social Media* (pp. 163–178). Hershey, PA: IGI Global. doi:10.4018/978-1-5225-0562-4.ch010

Butts, S. (2017). This Is the Remix: Remediating Pedagogy Practices. In K. Bryant (Ed.), *Engaging 21st Century Writers with Social Media* (pp. 74–95). Hershey, PA: IGI Global. doi:10.4018/978-1-5225-0562-4.ch005

Byrd, M. Y., & Lloyd-Jones, B. (2016). Developing a Social Justice-Oriented Workforce Diversity Concentration in Human Relations Academic Programs. In C. Scott & J. Sims (Eds.), *Developing Workforce Diversity Programs, Curriculum, and Degrees in Higher Education* (pp. 179–196). Hershey, PA: IGI Global. doi:10.4018/978-1-5225-0209-8.ch010

Callens, M. V. (2017). Using Wikipedia to Teach Written Health Communication. In K. Bryant (Ed.), *Engaging 21st Century Writers with Social Media* (pp. 247–258). Hershey, PA: IGI Global. doi:10.4018/978-1-5225-0562-4.ch015

Calvo, N., Bastida, M., & Feás, J. (2016). Is Gender Equality a Reality after a Change in the Political Agenda?: Dynamic Analysis for Higher Management Levels. In J. Prescott (Ed.), *Handbook of Research on Race, Gender, and the Fight for Equality* (pp. 539–558). Hershey, PA: IGI Global. doi:10.4018/978-1-5225-0047-6.ch023

Campbell, L. O., Truitt, J. H., Herlihy, C. P., & Plante, J. D. (2017). A Thematic Analysis of Leadership Qualities of Women Leaders in Technology: Viewed through Social Media. In V. Wang (Ed.), *Encyclopedia of Strategic Leadership and Management* (pp. 1–15). Hershey, PA: IGI Global. doi:10.4018/978-1-5225-1049-9.ch001

Carter, J. D. (2016). The African American Course Mystique: A Forum to Strengthen Student Engagement and Learning for White HBCU Students. In C. Prince, & R. Ford (Eds.), *Setting a New Agenda for Student Engagement and Retention in Historically Black Colleges and Universities* (pp. 186-221). Hershey, PA: IGI Global. doi:10.4018/978-1-5225-0308-8.ch011

Chapman, C. (2014). A Civic Engagement Graduation Requirement on an Urban College Campus. *International Journal of Civic Engagement and Social Change*, *1*(4), 1–27. doi:10.4018/IJCESC.2014100101

Cheikh-Ammar, M., & Barki, H. (2016). The Influence of Social Presence, Social Exchange and Feedback Features on SNS Continuous Use: The Facebook Context. *Journal of Organizational and End User Computing*, *28*(2), 33–52. doi:10.4018/JOEUC.2016040103

Cmeciu, C. (2016). Facebook Photographic Images: Political Tools of Self-Presentation during the 2014 European Parliament Elections in Romania. In T. Deželan & I. Vobič (Eds.), *Revolutionizing Political Communication through Social Media* (pp. 229–252). Hershey, PA: IGI Global. doi:10.4018/978-1-4666-9879-6.ch012

Cooper, L., Nash, M., & Umugwaneza, A. (2016). Social Work with Refugee Women: Narrative and "Empowerment". In J. Prescott (Ed.), *Handbook of Research on Race, Gender, and the Fight for Equality* (pp. 559–579). Hershey, PA: IGI Global. doi:10.4018/978-1-5225-0047-6.ch024

Cotton, K., Green, D. O., Beckman, S. A., Hussain, A., Robb, A., & Green, M. D. (2016). Leveraging New Media as Social Capital for Diversity Officers: How Equity, Diversity, and Inclusion Professionals Can Use Social Media to Foster Equality. In J. Prescott (Ed.), *Handbook of Research on Race, Gender, and the Fight for Equality* (pp. 113–141). Hershey, PA: IGI Global. doi:10.4018/978-1-5225-0047-6.ch006

Crosston, M. D. (2017). The Fight for Cyber Thoreau: Distinguishing Virtual Disobedience from Digital Destruction. In M. Korstanje (Ed.), *Threat Mitigation and Detection of Cyber Warfare and Terrorism Activities* (pp. 198–219). Hershey, PA: IGI Global. doi:10.4018/978-1-5225-1938-6.ch009

Dalisay, F., Kushin, M. J., & Yamamoto, M. (2016). Conflict as a Barrier to Online Political Participation?: A Look at Political Participation in an Era of Web and Mobile Connectivity. *International Journal of E-Politics*, *7*(1), 37–53. doi:10.4018/IJEP.2016010103

Dalisay, F., Kushin, M. J., & Yamamoto, M. (2017). The Demobilizing Potential of Conflict for Web and Mobile Political Participation. In Y. Ibrahim (Ed.), *Politics, Protest, and Empowerment in Digital Spaces* (pp. 52–71). Hershey, PA: IGI Global. doi:10.4018/978-1-5225-1862-4.ch004

Darling, J. (2017). From Expository Blog to Engaged E-Portfolio: A Student-Centered Pedagogy in Process. In K. Bryant (Ed.), *Engaging 21st Century Writers with Social Media* (pp. 51–72). Hershey, PA: IGI Global. doi:10.4018/978-1-5225-0562-4.ch004

del Mar Gálvez-Rodríhuez, M., Haro-de-Rosario, A., & Caba-Pérez, M. D. (2017). A Comparative View of Citizen Engagement in Social Media of Local Governments from North American Countries. In M. Adria & Y. Mao (Eds.), *Handbook of Research on Citizen Engagement and Public Participation in the Era of New Media* (pp. 139–156). Hershey, PA: IGI Global. doi:10.4018/978-1-5225-1081-9.ch009

DeMark, J., & Van Wingerden, C. (2017). Strategically Preparing the Next Generation for Leading in a World of Social Change. In V. Wang (Ed.), *Encyclopedia of Strategic Leadership and Management* (pp. 1719–1744). Hershey, PA: IGI Global. doi:10.4018/978-1-5225-1049-9.ch120

Demirhan, K. (2017). Participation, Civil Society and the Facebook Use of NGOs in Turkey. In K. Demirhan & D. Çakır-Demirhan (Eds.), *Political Scandal, Corruption, and Legitimacy in the Age of Social Media* (pp. 223–246). Hershey, PA: IGI Global. doi:10.4018/978-1-5225-2019-1.ch010

Demirhan, K. (2017). Scandal Politics and Political Scandals in the Era of Digital Interactive Media. In K. Demirhan & D. Çakır-Demirhan (Eds.), *Political Scandal, Corruption, and Legitimacy in the Age of Social Media* (pp. 25–50). Hershey, PA: IGI Global. doi:10.4018/978-1-5225-2019-1.ch002

Doughty, H. A. (2014). Critical Theory, Critical Pedagogy and the Permanent Crisis in Community Colleges. *International Journal of Adult Vocational Education and Technology*, *5*(2), 29–44. doi:10.4018/ijavet.2014040103

Drulă, G. (2017). Citizen Engagement and News Selection for Facebook Pages. In M. Adria & Y. Mao (Eds.), *Handbook of Research on Citizen Engagement and Public Participation in the Era of New Media* (pp. 122–138). Hershey, PA: IGI Global. doi:10.4018/978-1-5225-1081-9.ch008

During, R., Pleijte, M., van Dam, R. I., & Salverda, I. E. (2017). The Dutch Participation Society Needs Open Data, but What is Meant by Open? In M. Adria & Y. Mao (Eds.), *Handbook of Research on Citizen Engagement and Public Participation in the Era of New Media* (pp. 304–322). Hershey, PA: IGI Global. doi:10.4018/978-1-5225-1081-9.ch017

Eca, T. T., Saldanha, A., & Franco, A. M. (2017). "Insurgence": Activism in Art Education Research and Praxis. In R. Shin (Ed.), *Convergence of Contemporary Art, Visual Culture, and Global Civic Engagement* (pp. 210–223). Hershey, PA: IGI Global. doi:10.4018/978-1-5225-1665-1.ch012

Escolar, M. P. (2016). Imperfect Bipartisanship and Spanish Pluralism: The Keys to Success of Podemos on Twitter. In T. Deželan & I. Vobič (Eds.), *R)evolutionizing Political Communication through Social Media* (pp. 94–117). Hershey, PA: IGI Global. doi:10.4018/978-1-4666-9879-6.ch006

Essien, E. D. (2016). Ethical Appraisal of the Role of Civil Society in Nigeria: Pathway for Social Inclusion and Sustainable Development. *International Journal of Civic Engagement and Social Change*, *3*(2), 23–38. doi:10.4018/IJCESC.2016040102

Essien, E. D. (2017). Navigating the Nexus between Social Media, Political Scandal, and Good Governance in Nigeria: Its Ethical Implications. In K. Demirhan & D. Çakır-Demirhan (Eds.), *Political Scandal, Corruption, and Legitimacy in the Age of Social Media* (pp. 157–184). Hershey, PA: IGI Global. doi:10.4018/978-1-5225-2019-1.ch007

Falk, A. F., & Vine, B. E. (2017). Supporting Graduate Students to Implement Community-Engaged Research. In C. Crosby & F. Brockmeier (Eds.), *Student Experiences and Educational Outcomes in Community Engagement for the 21st Century* (pp. 196–224). Hershey, PA: IGI Global. doi:10.4018/978-1-5225-0874-8.ch009

Fiore, C. (2017). The Blogging Method: Improving Traditional Student Writing Practices. In K. Bryant (Ed.), *Engaging 21st Century Writers with Social Media* (pp. 179–198). Hershey, PA: IGI Global. doi:10.4018/978-1-5225-0562-4.ch011

Fredlund, K. (2017). Social Media and the Rhetorical Situation: Finding Common Ground Between Students' Lives and Writing Courses. In K. Bryant (Ed.), *Engaging 21st Century Writers with Social Media* (pp. 96–115). Hershey, PA: IGI Global. doi:10.4018/978-1-5225-0562-4.ch006

Gallardo, M., Jenkins, J. J., & Dillon, P. J. (2016). Exploring Identity Construction: Considerations of Student Identity Expression in Online Modalities. *International Journal of Information Communication Technologies and Human Development*, *8*(4), 32–37. doi:10.4018/IJICTHD.2016100103

Gambrell, J. A. (2017). Embracing Teaching as Social Activism: Rejecting Colorblindness through Cultural Inclusion during Study Abroad. In H. An (Ed.), *Handbook of Research on Efficacy and Implementation of Study Abroad Programs for P-12 Teachers* (pp. 322–340). Hershey, PA: IGI Global. doi:10.4018/978-1-5225-1057-4.ch018

Gessler, H. A. (2016). Reformulating Government-Citizen Relations in a Digitally Connected World: The Twitter Ban Phenomenon in Turkey. In T. Deželan & I. Vobič (Eds.), *R)evolutionizing Political Communication through Social Media* (pp. 75–93). Hershey, PA: IGI Global. doi:10.4018/978-1-4666-9879-6.ch005

Gismondi, A. (2014). OccupyWallStreet: Social Media, Education, and the Occupy Movement. In V. Benson & S. Morgan (Eds.), *Cutting-Edge Technologies and Social Media Use in Higher Education* (pp. 156–172). Hershey, PA: IGI Global. doi:10.4018/978-1-4666-5174-6.ch006

Gray, D. A. (2017). Art Empowering At-Risk and Opportunity Youth: A Case Study of artworxLA. In L. Hersey & B. Bobick (Eds.), *Handbook of Research on the Facilitation of Civic Engagement through Community Art* (pp. 320–338). Hershey, PA: IGI Global. doi:10.4018/978-1-5225-1727-6.ch015

Groshek, J. (2017). Organically Modified News Networks: Gatekeeping in Social Media Coverage of GMOs. In M. Adria & Y. Mao (Eds.), *Handbook of Research on Citizen Engagement and Public Participation in the Era of New Media* (pp. 107–121). Hershey, PA: IGI Global. doi:10.4018/978-1-5225-1081-9.ch007

Gupta, P. (2014). Feminist Jurisprudence in India with Reference to Individual Freedom of Women vis-a-vis State's Duty to Protect Them. *International Journal of Civic Engagement and Social Change*, *1*(2), 54–67. doi:10.4018/ijcesc.2014040104

Harrell, B. C. (2017). Using Twitter to Scaffold English Composition. In K. Bryant (Ed.), *Engaging 21st Century Writers with Social Media* (pp. 32–50). Hershey, PA: IGI Global. doi:10.4018/978-1-5225-0562-4.ch003

Hayes, K. (2017). Socializing Composition: Entering the Conversation of SNS in Composition. In K. Bryant (Ed.), *Engaging 21st Century Writers with Social Media* (pp. 133–150). Hershey, PA: IGI Global. doi:10.4018/978-1-5225-0562-4.ch008

Heaton, L., & Días da Silva, P. (2017). Citizen Engagement in Local Environmental Issues: Intersecting Modes of Communication. In M. Adria & Y. Mao (Eds.), *Handbook of Research on Citizen Engagement and Public Participation in the Era of New Media* (pp. 1–19). Hershey, PA: IGI Global. doi:10.4018/978-1-5225-1081-9.ch001

Heisler, E. (2017). Measuring Enclosures and Efficacy in Online Feminism: The Case of Rewire. In M. Adria & Y. Mao (Eds.), *Handbook of Research on Citizen Engagement and Public Participation in the Era of New Media* (pp. 389–409). Hershey, PA: IGI Global. doi:10.4018/978-1-5225-1081-9.ch021

Herkman, J., & Matikainen, J. (2017). Neo-Populist Scandal and Social Media: The Finnish Olli Immonen Affair. In K. Demirhan & D. Çakır-Demirhan (Eds.), *Political Scandal, Corruption, and Legitimacy in the Age of Social Media* (pp. 1–24). Hershey, PA: IGI Global. doi:10.4018/978-1-5225-2019-1.ch001

Hobbs, R. (2016). Lessons in Copyright Activism: K-12 Education and the DMCA 1201 Exemption Rulemaking Process. *International Journal of Information and Communication Technology Education, 12*(1), 50–63. doi:10.4018/IJICTE.2016010105

Holmes, J. W., & McNeal, R. S. (2016). Social Media, Participation, and Attitudes: Does Social Media Drive Polarization? In T. Deželan & I. Vobič (Eds.), *R)evolutionizing Political Communication through Social Media* (pp. 1–21). Hershey, PA: IGI Global. doi:10.4018/978-1-4666-9879-6.ch001

Huddleston, M. D. (2017). Impact of Community Engagement in Higher Education. In C. Crosby & F. Brockmeier (Eds.), *Student Experiences and Educational Outcomes in Community Engagement for the 21st Century* (pp. 1–28). Hershey, PA: IGI Global. doi:10.4018/978-1-5225-0874-8.ch001

Ifedi, R. (2017). African-Born Female Academics in the U.S.: Experiences of Inclusion, Exclusion, and Access - Building Careers on Marginalized Identities. *International Journal of Bias, Identity and Diversities in Education, 2*(1), 1–12. doi:10.4018/IJBIDE.2017010101

Iwasaki, Y. (2017). Youth Engagement in the Era of New Media. In M. Adria & Y. Mao (Eds.), *Handbook of Research on Citizen Engagement and Public Participation in the Era of New Media* (pp. 90–105). Hershey, PA: IGI Global. doi:10.4018/978-1-5225-1081-9.ch006

Jha, C. K. (2017). Information Control, Transparency, and Social Media: Implications for Corruption. In K. Demirhan & D. Çakır-Demirhan (Eds.), *Political Scandal, Corruption, and Legitimacy in the Age of Social Media* (pp. 51–75). Hershey, PA: IGI Global. doi:10.4018/978-1-5225-2019-1.ch003

Ji, Q. (2017). A Comparative View of Censored and Uncensored Political Discussion: The Case of Chinese Social Media Users. In M. Adria & Y. Mao (Eds.), *Handbook of Research on Citizen Engagement and Public Participation in the Era of New Media* (pp. 269–282). Hershey, PA: IGI Global. doi:10.4018/978-1-5225-1081-9.ch015

Jiang, S. (2016). "Microblog, A Public Sphere?": An Examination of Opinion Expression about "Snowden Affair" on Twitter and Sina Weibo. In T. Deželan & I. Vobič (Eds.), *R)evolutionizing Political Communication through Social Media* (pp. 22–39). Hershey, PA: IGI Global. doi:10.4018/978-1-4666-9879-6.ch002

Kaur, M., & Verma, R. (2016). Social Media: An Emerging Tool for Political Participation. *International Journal of Social and Organizational Dynamics in IT*, *5*(2), 31–38. doi:10.4018/IJSODIT.2016070103

Keating, P. (2016). Games for Social Change: Review of the Potential for Multi Player Online Gaming to Facilitate the Emergence and Growth of Progressive Movements for Social Change within Youth and Community Work. *International Journal of Game-Based Learning*, *6*(4), 73–86. doi:10.4018/IJGBL.2016100105

Keifer-Boyd, K. T. (2017). FemTechNet Distributed Open Collaborative Course: Performing Difference, Exquisite Engendering, and Feminist Mapping. In R. Shin (Ed.), *Convergence of Contemporary Art, Visual Culture, and Global Civic Engagement* (pp. 278–296). Hershey, PA: IGI Global. doi:10.4018/978-1-5225-1665-1.ch016

Kirschner, D. (2017). The Development of a Gamified System for Health Activism as a Graduate Student Project. In R. Zheng & M. Gardner (Eds.), *Handbook of Research on Serious Games for Educational Applications* (pp. 380–401). Hershey, PA: IGI Global. doi:10.4018/978-1-5225-0513-6.ch018

Korn, J. U. (2015). Black Nerds, Asian Activists, and Caucasian Dogs: Online Race-based Cultural Group Identities within Facebook Groups. *International Journal of Interactive Communication Systems and Technologies*, *5*(1), 14–25. doi:10.4018/IJICST.2015010102

Larsson, D., & Schmauch, U. (2017). Perception of Race Discrimination by the Police in Europe. In S. Egharevba (Ed.), *Police Brutality, Racial Profiling, and Discrimination in the Criminal Justice System* (pp. 13–37). Hershey, PA: IGI Global. doi:10.4018/978-1-5225-1088-8.ch002

Leurs, K. (2016). Young Connected Migrants and Non-Normative European Family Life: Exploring Affective Human Right Claims of Young E-Diasporas. *International Journal of E-Politics*, *7*(3), 15–34. doi:10.4018/IJEP.2016070102

López, C. M., Saborido, D. M., & Torrego, J. C. (2017). Detecting Problems of Coexistence: Giving Voice to Students. In S. Singh & N. Erbe (Eds.), *Creating a Sustainable Vision of Nonviolence in Schools and Society* (pp. 134–161). Hershey, PA: IGI Global. doi:10.4018/978-1-5225-2209-6.ch008

Maarhuis, P. L., & Rud, A. G. (2017). Dewey, School Violence, and Aesthetic Response: Healing the Community through Arts after Disaster. In L. Hersey & B. Bobick (Eds.), *Handbook of Research on the Facilitation of Civic Engagement through Community Art* (pp. 237–266). Hershey, PA: IGI Global. doi:10.4018/978-1-5225-1727-6.ch012

Maksuti, A., & Deželan, T. (2016). Post as Much and as Diversely as Possible: Intensity and Innovation in the Application of Twitter by Political Parties during the 2014 Slovenian National Election Campaign. In T. Deželan & I. Vobič (Eds.), *Revolutionizing Political Communication through Social Media* (pp. 168–189). Hershey, PA: IGI Global. doi:10.4018/978-1-4666-9879-6.ch009

Manguvo, A. (2016). Negotiating the Boundaries of American Blackness: The Experiences of African Students in the United States. In K. Bista & C. Foster (Eds.), *Exploring the Social and Academic Experiences of International Students in Higher Education Institutions* (pp. 79–95). Hershey, PA: IGI Global. doi:10.4018/978-1-4666-9749-2.ch005

Manrique, C. G., & Manrique, G. G. (2017). Social Media's Role in Alleviating Political Corruption and Scandals: The Philippines during and after the Marcos Regime. In K. Demirhan & D. Çakır-Demirhan (Eds.), *Political Scandal, Corruption, and Legitimacy in the Age of Social Media* (pp. 205–222). Hershey, PA: IGI Global. doi:10.4018/978-1-5225-2019-1.ch009

Marovitz, M. (2017). Social Networking Engagement and Crisis Communication Considerations. In M. Brown Sr., (Ed.), *Social Media Performance Evaluation and Success Measurements* (pp. 130–155). Hershey, PA: IGI Global. doi:10.4018/978-1-5225-1963-8.ch007

Martini, R. H., & Webster, T. (2017). What Online Writing Spaces Afford Us in the Age of Campus Carry, "Wall-Building," and Orlando's Pulse Tragedy. In E. Monske & K. Blair (Eds.), *Handbook of Research on Writing and Composing in the Age of MOOCs* (pp. 278–293). Hershey, PA: IGI Global. doi:10.4018/978-1-5225-1718-4.ch017

Maulana, I. (2017). Social Media as Public Political Instrument. In M. Adria & Y. Mao (Eds.), *Handbook of Research on Citizen Engagement and Public Participation in the Era of New Media* (pp. 231–247). Hershey, PA: IGI Global. doi:10.4018/978-1-5225-1081-9.ch013

Maxwell, S. P., & Carboni, J. L. (2017). Civic Engagement through Social Media: Strategic Stakeholder Management by High-Asset Foundations. *International Journal of Public Administration in the Digital Age*, *4*(1), 35–48. doi:10.4018/IJPADA.2017010103

McCarthy-Latimer, C., & Kendrick, J. R. Jr. (2016). How Communication Technologies Function as Platforms and Pathways to Civic and Political Engagement: An Agenda for Research and Public Policy. *International Journal of Civic Engagement and Social Change*, *3*(1), 50–72. doi:10.4018/IJCESC.2016010104

McGuire, M. (2017). Reblogging as Writing: The Role of Tumblr in the Writing Classroom. In K. Bryant (Ed.), *Engaging 21st Century Writers with Social Media* (pp. 116–131). Hershey, PA: IGI Global. doi:10.4018/978-1-5225-0562-4.ch007

McNeal, R. S., Schmeida, M., & Bryan, L. D. (2017). Smartphones and their Increased Importance in U.S. Presidential Elections. In M. Adria & Y. Mao (Eds.), *Handbook of Research on Citizen Engagement and Public Participation in the Era of New Media* (pp. 283–303). Hershey, PA: IGI Global. doi:10.4018/978-1-5225-1081-9.ch016

Mhiripiri, N. A., & Ureke, O. (2016). Mobile Journalism, Cellphilms, and the Use of the StoryMaker Multimedia Software at a Zimbabwean Media Training University. In L. Briz-Ponce, J. Juanes-Méndez, & F. García-Peñalvo (Eds.), *Handbook of Research on Mobile Devices and Applications in Higher Education Settings* (pp. 318–343). Hershey, PA: IGI Global. doi:10.4018/978-1-5225-0256-2.ch014

Mostafa, R. H., & Attia, S. T. (2015). The Effect of Social Networks Sites (SNSs) on the Egyptian 25/30 Uprisings. *International Journal of Online Marketing*, *5*(2), 58–74. doi:10.4018/IJOM.2015040104

Muñoz, L. A., Bolívar, M. P., & Quiles, F. J. (2016). Policies and Strategies for Digital Inclusion: Regional Governments in Spain. In J. Prescott (Ed.), *Handbook of Research on Race, Gender, and the Fight for Equality* (pp. 1–29). Hershey, PA: IGI Global. doi:10.4018/978-1-5225-0047-6.ch001

Murru, M. F. (2016). The Performative Role of Hashtags in the Politicization of Europe: Twitter and European Elections in 2014. In T. Deželan & I. Vobič (Eds.), *R)evolutionizing Political Communication through Social Media* (pp. 147–167). Hershey, PA: IGI Global. doi:10.4018/978-1-4666-9879-6.ch008

Neimann, T. D., & Stelson, U. M. (2017). Prevalence of Campus Rape: Strategic Leadership Prevention and Recommendations. In V. Wang (Ed.), *Encyclopedia of Strategic Leadership and Management* (pp. 1427–1449). Hershey, PA: IGI Global. doi:10.4018/978-1-5225-1049-9.ch099

Noelk, D. V. (2017). Effecting Change in a Politically Charged Environment. In V. Wang (Ed.), *Encyclopedia of Strategic Leadership and Management* (pp. 823–832). Hershey, PA: IGI Global. doi:10.4018/978-1-5225-1049-9.ch057

Noor, R. (2017). Citizen Journalism: News Gathering by Amateurs. In M. Adria & Y. Mao (Eds.), *Handbook of Research on Citizen Engagement and Public Participation in the Era of New Media* (pp. 194–229). Hershey, PA: IGI Global. doi:10.4018/978-1-5225-1081-9.ch012

Novoselova, V. (2016). Digitizing Consumer Activism: A Thematic Analysis of Jezebel.com. In A. Novak & I. El-Burki (Eds.), *Defining Identity and the Changing Scope of Culture in the Digital Age* (pp. 140–154). Hershey, PA: IGI Global. doi:10.4018/978-1-5225-0212-8.ch009

Oluwole, J. O., & Green, P. C. III. (2016). Acceptable Use Policies. In *Censorship and Student Communication in Online and Offline Settings* (pp. 352–358). Hershey, PA: IGI Global. doi:10.4018/978-1-4666-9519-1.ch014

Oluwole, J. O., & Green, P. C. III. (2016). Advocacy of Illegal Drug Use. In *Censorship and Student Communication in Online and Offline Settings* (pp. 108–121). Hershey, PA: IGI Global. doi:10.4018/978-1-4666-9519-1.ch010

Oluwole, J. O., & Green, P. C. III. (2016). Assessing the Current Jurisprudence. In *Censorship and Student Communication in Online and Offline Settings* (pp. 304–351). Hershey, PA: IGI Global. doi:10.4018/978-1-4666-9519-1.ch013

Oluwole, J. O., & Green, P. C. III. (2016). Background. In *Censorship and Student Communication in Online and Offline Settings* (pp. 1–18). Hershey, PA: IGI Global. doi:10.4018/978-1-4666-9519-1.ch001

Oluwole, J. O., & Green, P. C. III. (2016). Child Pornography. In *Censorship and Student Communication in Online and Offline Settings* (pp. 55–60). Hershey, PA: IGI Global. doi:10.4018/978-1-4666-9519-1.ch006

Oluwole, J. O., & Green, P. C. III. (2016). Defamatory Speech. In *Censorship and Student Communication in Online and Offline Settings* (pp. 44–49). Hershey, PA: IGI Global. doi:10.4018/978-1-4666-9519-1.ch004

Oluwole, J. O., & Green, P. C. III. (2016). Fighting Words. In *Censorship and Student Communication in Online and Offline Settings* (pp. 36–43). Hershey, PA: IGI Global. doi:10.4018/978-1-4666-9519-1.ch003

Oluwole, J. O., & Green, P. C. III. (2016). Forging the Path Forward from Censorship. In *Censorship and Student Communication in Online and Offline Settings* (pp. 590–605). Hershey, PA: IGI Global. doi:10.4018/978-1-4666-9519-1.ch016

Oluwole, J. O., & Green, P. C. III. (2016). Lewd, Vulgar, Plainly-Offensive, and Obscene Speech. In *Censorship and Student Communication in Online and Offline Settings* (pp. 84–96). Hershey, PA: IGI Global. doi:10.4018/978-1-4666-9519-1.ch008

Oluwole, J. O., & Green, P. C. III. (2016). Material and Substantial Disruption/Infringement of Rights. In *Censorship and Student Communication in Online and Offline Settings* (pp. 62–83). Hershey, PA: IGI Global. doi:10.4018/978-1-4666-9519-1.ch007

Oluwole, J. O., & Green, P. C. III. (2016). Obscene Speech. In *Censorship and Student Communication in Online and Offline Settings* (pp. 50–54). Hershey, PA: IGI Global. doi:10.4018/978-1-4666-9519-1.ch005

Oluwole, J. O., & Green, P. C. III. (2016). School-Sponsored Speech. In *Censorship and Student Communication in Online and Offline Settings* (pp. 97–107). Hershey, PA: IGI Global. doi:10.4018/978-1-4666-9519-1.ch009

Oluwole, J. O., & Green, P. C. III. (2016). State Anti-Bullying Statutes and Student Speech. In *Censorship and Student Communication in Online and Offline Settings* (pp. 359–589). Hershey, PA: IGI Global. doi:10.4018/978-1-4666-9519-1.ch015

Oluwole, J. O., & Green, P. C. III. (2016). Their Stories: Offline Off-Campus Speech. In *Censorship and Student Communication in Online and Offline Settings* (pp. 123–207). Hershey, PA: IGI Global. doi:10.4018/978-1-4666-9519-1.ch011

Oluwole, J. O., & Green, P. C. III. (2016). True Threat. In *Censorship and Student Communication in Online and Offline Settings* (pp. 20–35). Hershey, PA: IGI Global. doi:10.4018/978-1-4666-9519-1.ch002

Oravec, J. A. (2014). Mottos and Ethical Statements of Internet-Based Organizations: Implications for Corporate Social Responsibility. *International Journal of Civic Engagement and Social Change*, *1*(2), 37–53. doi:10.4018/ijcesc.2014040103

Overly, D. (2017). Saving Civilization: Strategies for Peace. In S. Singh & N. Erbe (Eds.), *Creating a Sustainable Vision of Nonviolence in Schools and Society* (pp. 260–277). Hershey, PA: IGI Global. doi:10.4018/978-1-5225-2209-6.ch014

Papagiannidis, S., & Manika, D. (2016). Political Participation and Engagement via Different Online and Offline Channels. *International Journal of E-Business Research*, *12*(4), 1–22. doi:10.4018/IJEBR.2016100101

Papasolomou, I. (2017). The Concept of "Consumerism" from a Consumer Activist Perspective. In H. Kaufmann & M. Panni (Eds.), *Socio-Economic Perspectives on Consumer Engagement and Buying Behavior* (pp. 1–21). Hershey, PA: IGI Global. doi:10.4018/978-1-5225-2139-6.ch001

Perelló-Sobrepere, M. (2017). Building a New State from Outrage: The Case of Catalonia. In M. Adria & Y. Mao (Eds.), *Handbook of Research on Citizen Engagement and Public Participation in the Era of New Media* (pp. 344–359). Hershey, PA: IGI Global. doi:10.4018/978-1-5225-1081-9.ch019

Phillips, S. N. (2017). Networked Digital Spaces: Twitter in the Composition Classroom. In K. Bryant (Ed.), *Engaging 21st Century Writers with Social Media* (pp. 1–13). Hershey, PA: IGI Global. doi:10.4018/978-1-5225-0562-4.ch001

Poulakidakos, S., & Veneti, A. (2016). Political Communication and Twitter in Greece: Jumps on the Bandwagon or an Enhancement of the Political Dialogue? In T. Deželan & I. Vobič (Eds.), *R)evolutionizing Political Communication through Social Media* (pp. 119–146). Hershey, PA: IGI Global. doi:10.4018/978-1-4666-9879-6.ch007

Ramluckan, T., Ally, S. E., & van Niekerk, B. (2017). Twitter Use in Student Protests: The Case of South Africa's #FeesMustFall Campaign. In M. Korstanje (Ed.), *Threat Mitigation and Detection of Cyber Warfare and Terrorism Activities* (pp. 220–253). Hershey, PA: IGI Global. doi:10.4018/978-1-5225-1938-6.ch010

Rathnayake, C. (2015). From Street Protests to Facebook Campaigns: Political Cynicism, Efficacy and Online Political Engagement of Sri Lankan Students. *International Journal of E-Politics, 6*(1), 42–58. doi:10.4018/IJEP.2015010104

Roth, K. R., & Ritter, Z. S. (2016). Channeling Race: Media Representations and International Student Perceptions. In J. Prescott (Ed.), *Handbook of Research on Race, Gender, and the Fight for Equality* (pp. 172–197). Hershey, PA: IGI Global. doi:10.4018/978-1-5225-0047-6.ch008

Roth, K. R., & Ritter, Z. S. (2016). Racial Spectacle and Campus Climate: Media Representations and Asian International Student Perceptions at U.S. Colleges. In J. Prescott (Ed.), *Handbook of Research on Race, Gender, and the Fight for Equality* (pp. 142–171). Hershey, PA: IGI Global. doi:10.4018/978-1-5225-0047-6.ch007

Rubens, A. (2017). Teaching Casual Writing for Professional Success with Twitter: Digital Small Talk and the New Textese. In K. Bryant (Ed.), *Engaging 21st Century Writers with Social Media* (pp. 200–220). Hershey, PA: IGI Global. doi:10.4018/978-1-5225-0562-4.ch012

Rudge, L. T. (2017). Different Experiences and Perceptions of Campus Climate Among Minority Students at a Predominantly White Institution. *International Journal of Bias, Identity and Diversities in Education*, *2*(1), 40–56. doi:10.4018/IJBIDE.2017010104

Sandoval-Almazan, R. (2017). Political Messaging in Digital Spaces: The Case of Twitter in Mexico's Presidential Campaign. In Y. Ibrahim (Ed.), *Politics, Protest, and Empowerment in Digital Spaces* (pp. 72–90). Hershey, PA: IGI Global. doi:10.4018/978-1-5225-1862-4.ch005

Santamaría, L. J. (2015). Culturally Responsive Educational Leadership in Cross-Cultural International Contexts. In N. Erbe & A. Normore (Eds.), *Cross-Cultural Collaboration and Leadership in Modern Organizations* (pp. 120–139). Hershey, PA: IGI Global. doi:10.4018/978-1-4666-8376-1.ch008

Sebubi, O. (2016). Voters and Mobile: Impact on Democratic Revolution. In X. Xu (Ed.), *Handbook of Research on Human Social Interaction in the Age of Mobile Devices* (pp. 150–170). Hershey, PA: IGI Global. doi:10.4018/978-1-5225-0469-6.ch008

Seifert, J. L. (2015). A Culture of Survivors: SlutWalk, Third Culture, and New Media Communication. *International Journal of Interactive Communication Systems and Technologies*, *5*(1), 1–13. doi:10.4018/IJICST.2015010101

Shafi, A., & Vultee, F. (2016). One of Many Tools to Win the Election: A Study of Facebook Posts by Presidential Candidates in the 2012 Election. In T. Deželan & I. Vobič (Eds.), *R)evolutionizing Political Communication through Social Media* (pp. 210–228). Hershey, PA: IGI Global. doi:10.4018/978-1-4666-9879-6.ch011

Shan, Z., & Tang, L. (2017). Social Media and the Public Sphere in China: A Case Study of Political Discussion on Weibo after the Wenzhou High-Speed Rail Derailment Accident. In M. Adria & Y. Mao (Eds.), *Handbook of Research on Citizen Engagement and Public Participation in the Era of New Media* (pp. 410–425). Hershey, PA: IGI Global. doi:10.4018/978-1-5225-1081-9.ch022

Song, M. Y., & Abelson, J. (2017). Public Engagement and Policy Entrepreneurship on Social Media in the Time of Anti-Vaccination Movements. In M. Adria & Y. Mao (Eds.), *Handbook of Research on Citizen Engagement and Public Participation in the Era of New Media* (pp. 38–56). Hershey, PA: IGI Global. doi:10.4018/978-1-5225-1081-9.ch003

Spada, P., & Allegretti, G. (2017). Integrating Multiple Channels of Engagement in Democratic Innovations: Opportunities and Challenges. In M. Adria & Y. Mao (Eds.), *Handbook of Research on Citizen Engagement and Public Participation in the Era of New Media* (pp. 20–37). Hershey, PA: IGI Global. doi:10.4018/978-1-5225-1081-9.ch002

Stacey, E. (2015). Networked Protests: A Review of Social Movement Literature and the Hong Kong Umbrella Movement. *International Journal of Civic Engagement and Social Change*, *2*(3), 36–51. doi:10.4018/IJCESC.2015070103

Strekalova, Y. A., Krieger, J. L., Damiani, R. E., Kalyanaraman, S., & Wang, D. Z. (2017). Old Media, New Media, and Public Engagement with Science and Technology. In M. Adria & Y. Mao (Eds.), *Handbook of Research on Citizen Engagement and Public Participation in the Era of New Media* (pp. 57–72). Hershey, PA: IGI Global. doi:10.4018/978-1-5225-1081-9.ch004

Suki, N. M. (2015). Engagement in Campus Environmental Activities and Green Initiatives in Malaysia: A Structural Equation Modelling Approach. *International Journal of Social Ecology and Sustainable Development*, *6*(4), 15–33. doi:10.4018/IJSESD.2015100102

Toscano, J. P. (2017). Social Media and Public Participation: Opportunities, Barriers, and a New Framework. In M. Adria & Y. Mao (Eds.), *Handbook of Research on Citizen Engagement and Public Participation in the Era of New Media* (pp. 73–89). Hershey, PA: IGI Global. doi:10.4018/978-1-5225-1081-9.ch005

Towner, T. L. (2017). Information Hubs or Drains?: The Role of Online Sources in Campaign Learning. In M. Adria & Y. Mao (Eds.), *Handbook of Research on Citizen Engagement and Public Participation in the Era of New Media* (pp. 157–178). Hershey, PA: IGI Global. doi:10.4018/978-1-5225-1081-9.ch010

Trauth, E. (2017). Creating Meaning for Millennials: Bakhtin, Rosenblatt, and the Use of Social Media in the Composition Classroom. In K. Bryant (Ed.), *Engaging 21st Century Writers with Social Media* (pp. 151–162). Hershey, PA: IGI Global. doi:10.4018/978-1-5225-0562-4.ch009

Trauth, E. (2017). Curating the Public Self: Helping Students Present an Authentic, Professional Persona via LinkedIn. In K. Bryant (Ed.), *Engaging 21st Century Writers with Social Media* (pp. 221–230). Hershey, PA: IGI Global. doi:10.4018/978-1-5225-0562-4.ch013

Tynes, B. M., Rose, C. A., Hiss, S., Umaña-Taylor, A. J., Mitchell, K., & Williams, D. (2014). Virtual Environments, Online Racial Discrimination, and Adjustment among a Diverse, School-Based Sample of Adolescents. *International Journal of Gaming and Computer-Mediated Simulations*, *6*(3), 1–16. doi:10.4018/ijgcms.2014070101 PMID:27134698

Utman, J. S. (2015). No, it did Not Grow Up because of the Internet: The Emergence of 2011's Student Mobilization in Chile. *International Journal of E-Politics*, *6*(4), 35–52. doi:10.4018/IJEP.2015100103

Vatikiotis, P. (2016). Social Media Activism: A Contested Field. In T. Deželan & I. Vobič (Eds.), *R)evolutionizing Political Communication through Social Media* (pp. 40–54). Hershey, PA: IGI Global. doi:10.4018/978-1-4666-9879-6.ch003

Vesnic-Alujevic, L. (2016). European Elections and Facebook: Political Advertising and Deliberation? In T. Deželan & I. Vobič (Eds.), *R)evolutionizing Political Communication through Social Media* (pp. 191–209). Hershey, PA: IGI Global. doi:10.4018/978-1-4666-9879-6.ch010

Vie, S., Carter, D., & Meyr, J. (2017). Occupy Rhetoric: Responding to Charges of "Slacktivism" with Digital Activism Successes. In M. Adria & Y. Mao (Eds.), *Handbook of Research on Citizen Engagement and Public Participation in the Era of New Media* (pp. 179–193). Hershey, PA: IGI Global. doi:10.4018/978-1-5225-1081-9.ch011

Walker, C. J. (2017). Blog Love: Blogging (And Microblogging) Communities as Writing Classroom Companions. In K. Bryant (Ed.), *Engaging 21st Century Writers with Social Media* (pp. 14–31). Hershey, PA: IGI Global. doi:10.4018/978-1-5225-0562-4.ch002

Webster, N., Coffey, H., & Ash, A. (2016). UrbanLivesMatter: Empowering Learners through Transformative Teaching. In T. Petty, A. Good, & S. Putman (Eds.), *Handbook of Research on Professional Development for Quality Teaching and Learning* (pp. 462–485). Hershey, PA: IGI Global. doi:10.4018/978-1-5225-0204-3.ch022

Willems, L. G., & Alizadeh, T. (2015). Social Media for Public Involvement and Sustainability in International Planning and Development. *International Journal of E-Planning Research*, *4*(4), 1–17. doi:10.4018/ijepr.2015100101

Wohn, D. Y., Lee, Y., & Ozkaya, E. Y. (2015). Social Contributors and Consequences of Habitual and Compulsive Game Play. *International Journal of Technology and Human Interaction*, *11*(3), 17–34. doi:10.4018/ijthi.2015070102

Wright, G. (2017). How to Create Sustainable Nonviolence Curriculum in Public Schools in Sweden: A Case Study. In S. Singh & N. Erbe (Eds.), *Creating a Sustainable Vision of Nonviolence in Schools and Society* (pp. 162–181). Hershey, PA: IGI Global. doi:10.4018/978-1-5225-2209-6.ch009

Xiaoming, H., Nainan, W., & George, C. (2014). The Impact of Online News Consumption on Young People's Political Participation. *International Journal of E-Politics*, *5*(2), 16–31. doi:10.4018/ijep.2014040102

Xu, Q., & Qi, L. (2014). Use of SNSs, Political Efficacy, and Civic Engagement among Chinese College Students: Effects of Gratifications and Network Size. *International Journal of Interactive Communication Systems and Technologies*, *4*(1), 15–30. doi:10.4018/ijicst.2014010102

Yang, K. C., & Kang, Y. (2017). Social Media, Political Mobilization, and Citizen Engagement: A Case Study of the March 18, 2014, Sunflower Student Movement in Taiwan. In M. Adria & Y. Mao (Eds.), *Handbook of Research on Citizen Engagement and Public Participation in the Era of New Media* (pp. 360–388). Hershey, PA: IGI Global. doi:10.4018/978-1-5225-1081-9.ch020

Yefet, B. (2017). The Clash of Civil Religions in Post-Revolutionary Egypt. In E. Lewin, E. Bick, & D. Naor (Eds.), *Comparative Perspectives on Civil Religion, Nationalism, and Political Influence* (pp. 156–189). Hershey, PA: IGI Global. doi:10.4018/978-1-5225-0516-7.ch006

Yuan, Y. (2017). A Gradual Political Change?: The Agenda Setting Effect of Online Activism in China 1994-2011. In M. Adria & Y. Mao (Eds.), *Handbook of Research on Citizen Engagement and Public Participation in the Era of New Media* (pp. 248–268). Hershey, PA: IGI Global. doi:10.4018/978-1-5225-1081-9.ch014

Yuce, S. T., Agarwal, N., Wigand, R. T., Lim, M., & Robinson, R. S. (2014). Bridging Women Rights Networks: Analyzing Interconnected Online Collective Actions. *Journal of Global Information Management*, *22*(4), 1–20. doi:10.4018/jgim.2014100101

Index

A

App 55, 64, 80-81
Authoring 34, 47, 51, 53, 62
Awareness 1, 37, 101, 105, 113, 125, 140

C

Children 7, 13-19, 27, 29, 34-35, 37-38, 45-51, 57, 59-60, 85, 96, 105, 112, 114, 123, 125, 128, 132-134, 138, 141, 143
Choices 23, 30, 37, 73, 98, 120, 136, 140, 146
Citizenship 38, 84, 123, 140-141
Cloud 81-82
Confusion 45, 54-55, 57-58, 73, 75, 79, 81-82, 111, 124, 132
Connection 17, 63, 73, 101-102, 108-109, 113, 128, 136
Consumerism 7, 117
Cooperation 15, 54
Creativity 13, 51, 74, 97-98
Culture 3-5, 7, 9, 11, 14-15, 35, 38, 42, 45-46, 51, 61, 74-75, 83-86, 94, 96, 98, 101, 116, 119-120, 126, 129

D

Data 9, 15, 27, 32, 39, 50, 69-70, 74, 81-82, 88, 98, 102, 107, 109, 123, 125, 140
Degrowth 148
devices 1-2, 5, 9-11, 23-26, 30-31, 38, 40, 42, 45-48, 52-54, 58, 60-61, 66, 68-69, 74-75, 77-78, 80-85, 90-91, 94, 96, 102, 105, 107-108, 114, 119, 126-127, 134, 138, 140
Digital 1-2, 5, 9-11, 24, 31, 33-38, 42-43, 45-46, 49-55, 58, 60, 64-65, 67-70, 73, 75-77, 80, 82-86, 90, 93-97, 105, 107-110, 113, 116-117, 124, 135, 138, 140-141, 147

E

Economy 74, 79-80, 98, 123-124, 134, 143, 145-146, 148
Education 13-14, 17-19, 24, 34, 46, 54, 56-57, 61-62, 64-65, 106, 111, 125, 138, 140, 143
Ethic 116, 134, 147
Example 3, 10-11, 14-15, 24, 35-37, 39, 41, 49-50, 53, 58, 61, 70, 76, 78, 82, 85, 89, 95, 104-105, 113, 117, 119, 123-124, 126, 129-132, 134-135, 137, 139-142, 144, 146

F

Food 101, 119-121, 126, 135, 146
Future 1, 6-7, 9, 13, 36, 38-39, 41, 50, 52, 55, 66, 68, 70, 73-74, 76-78, 80, 86, 94-96, 105, 112, 114-115, 119-120, 124-125, 136, 145-148

G

Generations 2-3, 10, 12, 28, 46, 49, 68-69, 83, 94, 102, 104, 135, 141
Google 82, 90-91, 94, 121, 135-136

H

Hacker 74, 116, 134, 143
History 2, 19, 26, 29, 42, 51, 88, 105, 120, 126, 129, 134, 143, 146

I

Ideology 1, 24, 28-29, 38, 41, 70, 117
Illich 24, 32, 41
Intelligence 15, 38, 68, 112, 114

K

Kids 7, 14-17, 19, 31-32, 34, 37, 45-49, 51-53, 57-65, 83, 110, 112-114, 119, 125-126, 128, 130-134, 141-142

L

Literacy 18, 33, 53-54, 58-59, 61, 68, 80-81, 84-85, 93

M

Maps 26, 90, 135-138, 140, 144
Marcuse 7-8, 13
Market 1-2, 7, 9-11, 17, 23-24, 26, 28-31, 38, 40-43, 52, 57, 66, 73, 75-77, 80, 89-91, 95, 101, 105, 108, 111, 115-116, 119-121, 124, 127, 134, 145, 147
McLuhan 8-9, 56
Movement 4-6, 13, 42, 47, 58, 69, 74, 94, 112, 114, 138, 148
Multimedia 23, 28, 45-47, 51-53, 55, 57, 62, 92, 107, 128
Music 2-3, 6-7, 10-11, 28, 47, 49, 54, 57, 60, 63, 68, 73, 76-77, 92, 97, 102, 107-108, 110, 118-119, 125
Myth 8, 74

N

newspapers 10, 38, 45

P

Participation 4, 75, 81, 98, 101, 104, 123, 137, 140, 144
Personal Computer 26-27, 41, 51, 102
Philosophy 141-143
Photography 125, 135
Post-Truth 75, 79-80, 84
Pretending 50-51, 59
Production 17, 23, 32, 42-43, 55, 60, 76, 101, 110-111, 119-121, 123, 129, 135, 140, 144-147
Programming language 18, 33, 112

R

Robots 9, 68, 112-113

S

Science 9-10, 28, 51, 84, 125-127, 131, 137
Senses 15, 38, 54, 56, 85, 106, 113, 116-117
Sharing 5, 17, 39-40, 42, 52, 91-93, 97-98, 123-125, 128-129, 134-135, 145-148
Sixties 2-4, 7-8, 10-11, 14, 16, 18, 50

Social Networks 7, 10, 49, 54, 58, 73-74, 76, 79, 81, 84, 93, 95, 97-98, 101, 104, 113
Software 27-28, 31-32, 34, 37, 41-42, 45, 47, 50-54, 63-64, 66, 68, 70, 73, 79, 81, 84, 86, 95-97, 101, 105, 107, 113, 121, 124-125, 127, 134-136, 138, 144-145, 147
Students 4, 6, 11, 14, 16-17, 23, 25-26, 40, 42, 49, 52, 60, 65, 67, 137-138, 141

T

Television 25, 45, 47, 55-56, 58, 61, 74-75, 84, 102, 108-109, 116-117
Tools 19, 25, 32, 38-39, 41, 52, 55, 57, 60-61, 63, 85, 91, 95-96, 105, 107, 123, 125-126, 128, 137, 140

V

Video 5, 10, 23-30, 32, 46-48, 50-53, 55-57, 60-63, 65-66, 69, 76-79, 82, 91-92, 94-95, 97, 102-103, 106-108, 112, 116, 119, 125-128, 132-134, 137, 142

W

War 2, 4, 7-8, 95, 98, 115, 119, 140
Web 41-42, 49, 52, 54-55, 57, 70, 74, 79-84, 92-93, 95, 97, 101-105, 107, 125, 127-129, 134, 136-138, 144
Women 3, 5-6, 14, 19, 32, 37-38, 58, 119, 140

www.ingramcontent.com/pod-product-compliance
Lightning Source LLC
LaVergne TN
LVHW062125200926
848484LV00007B/1764

* 9 7 8 1 5 2 2 5 3 7 7 0 0 *